ELEMENTA

Journal of Slavic Studies and Comparative Cultural Semiotics

Vol. 1, No. 1

1993

Vyacheslav Ivanov, Editor
University of California, Los Angeles
Moscow University

Harwood Academic Publishers

Switzerland • Australia • Belgium • France • Germany • Great Britain
India • Japan • Malaysia • Netherlands • Russia • Singapore • USA

AIMS AND SCOPE

ELEMENTA
Journal of Slavic Studies and Comparative Cultural Semiotics

- proposes the development of a new branch of semiotics: the science of cultural symbols and texts in its relationship to Slavic studies;
- continues and develops investigations begun by the Prague circle, the Moscow–Tartu school of semiotics and their predecessors;
- publishes articles on general problems of sign systems; evaluates the results of the comparison of different semiotic systems and texts in their relationship to Slavic traditions; opens up new archival materials that bear on the history of semiotic systems and semiotic historiography; and includes "miscellanea," i.e., short notes and comments to various texts;
- emphasizes the semiotic legacy of those scholars from different Slavic and other Eastern and Central European countries whose work is not yet sufficiently known to a Western audience due to linguistic barriers and historical obstacles.

ELEMENTA gives special attention to:

- General problems of sign and culture
- Semiotics of texts
- Slavic Avant-Garde
- Russian cultural history
- Slavs and other peoples of Eurasia
- Slavic archival materials
- Topical themes highlighting:
  - single writers
  - influential scholars
  - important semiotic problems.

Invitation to Contribute

Submissions may fall in the following categories: articles, review articles, short notes and comments. A companion monograph series will include original scholarly studies and new archival materials, particularly drawn from the heritage of those Slavic scholars whose works were not published during their lifetime.

Notes for contributors can be found at the back of the journal.

Distributed for Harwood Academic Publishers GmbH by STBS Ltd, PO Box 90, Reading, Berkshire RG1 8JL, Great Britain.

Printed in the United States of America

JANUARY 1993

Contents

ABOUT THE CONTRIBUTORS

Vyacheslav Ivanov is Professor of Slavic Languages and Literatures at the University of California, Los Angeles; Chair of the Department of the History and Theory of World Culture, Moscow University; and Director of the Institute of World Culture of Moscow University. He is a corresponding member of the British Academy.

He was one of the founders of the Moscow–Tartu school of semiotics; and has published numerous monographs and articles on semiotics of different sign systems and texts (natural and artificial languages, visual art, cinema, literature, religion) and on the history of semiotics (the book *Studies on the History of Russian Semiotics*; articles on Bakhtin, Vygotsky, Eisenstein, Freidenberg and other scholars).

He has been exploring problems of neurosemiotics (aphasia, functions of the two hemispheres of the brain in connection with various components of language and music); he published the book *The Odd and the Even: Asymmetry of the Brain and the Sign Systems* (translated into Japanese, German and other languages). His current research is concentrated on the poetics of Pasternak and Mandelstam.

Henrik Birnbaum is Professor of Slavic Languages and Literatures at the University of California, Los Angeles. He is the author of numerous books, monographs, and essays in Slavic, Balkan, and Indo-European linguistics; language typology, linguistic theory; early Slavic civilization; medieval Russian history; and modern Russian and German literature. He is a member of the American Academy of Arts and Sciences, and a corresponding member of the Swedish, Croatian, and Polish Academies of Sciences.

Carlo Ginzburg is a specialist on the history of culture and consciousness of early modern Europe, and is the author of numerous studies on the methodology of history. His books in English include: *The Cheese and the Worms: The Cosmos of 16th Century Miller* (1980); *The Night Battles: Witchcraft and Agrarian Cults in the 16th and 17th Centuries* (1983); *The Enigma of Piero* (1985); *Clues, Emblems and Myths* (1989); and *Ecstasies: Deciphering the Witches' Sabbath* (1991).

Omry Ronen teaches Russian literature at the University of Michigan, Ann Arbor. He is the author of *An Approach to Mandelstam* (Jerusalem, 1983), and numerous explications of Russian poetry and studies in poetics, especially the theory of subtext. Currently he is working on a monograph entitled *The Fallacy of the Silver Age*.

Yelena Dushechkina is a specialist on Russian literature and folklore. Her publications include a monograph on the 17th-century "Tale of Frol Skobeev," as well as articles on medieval literature, and on 18th–19th century poetry and prose. Her current research and writing has centered on calendar prose, in its historical development and in relationship to its cultural milieu; she has also edited several anthologies of Yuletide tales. She presently teaches at St. Petersburg University.

Henryk Baran writes on 19th–20th century Russian literature, poetics and semiotics. His current interests focus on the relationship of literature to myth and folklore, on the one hand, and to the press and popular culture, on the other. He recently edited a volume of articles by the late Krystyna Pomorska, *Jakobsonian Poetics and Slavic Narrative: From Pushkin to Solzhenitsyn* (1992). A selection of his papers will appear in 1993 in Moscow.

Alexander Ospovat is the author of books on the poet Fedor Tiutchev (1980) and Pushkin's poem *The Bronze Horseman* (with Roman Timenchik—1985, 2nd ed. 1987), as well as articles on the history of Russian literature. His current research and writing focuses on problems of culture and intellectual history in Russia of the first half of the 19th century.

Mikhail Yampolsky, a senior research fellow of the Institute of Philosophy of the Russian Academy of Sciences, is the author of books *Memory-Tiresias* (1992) and *Visible World* (1992), and of articles on problems of film art and 18th–20th century European culture. Currently he is Visiting Professor at New York University.

EDITORIAL

The scope of *Elementa: Journal of Slavic Studies and Comparative Cultural Semiotics* includes both comparative research on various aspects of modern cross-cultural sign systems and the results of the application of this approach to different Slavic traditions. The aim of the journal is to continue and develop waves of investigation begun by the Prague Circle and the Moscow–Tartu school of semiotics as well as their predecessors (particularly those scholars from East and Central Europe whose works are not sufficiently known to a Western audience due to linguistic barriers and historical obstacles). Moreover, a new synthesis of these semiotic trends with Saussurean and Peircean traditions will be attempted. Some general schemes will be elaborated to serve as tools for the study of different fields of intellectual activity expressed in verbal and non-verbal texts as symbols, icons and indexes. Among general approaches to religion, mythology, ritual, literature, and the visual arts, particular attention must be paid to a new type of model according to which they may be seen as special kinds of signs, sign systems and texts.

This semiotic approach may be termed linguistic in a figurative sense since we consider each of these fields specific "languages" with their own syntactic, semantic and pragmatic structure; internal relations; and with the possibility of describing some fragments of the universe, of society, and of the individual. Yet it is not possible to rely on strictly linguistic models to analyze verbal and non-verbal signs and texts. The restricted set of elements and rules of their combination developed for such discrete texts as those of folklore is not an adequate model for continuous forms of cultural transmission of information. Language proves to be not the only primary modeling system. Thus, it is not only possible to see the limitations of linguistic models as applied, for example, to music or cinema, but it also proves important to study the possibilities of cinema-like models of poetic language. An example is the application of the montage technique of Sergei Eisenstein's cinematographic theory to poetics. The broader semantic approach advocated by contributors to *Elementa* will therefore be free from purely linguistic limitations.

In the ancient world, as well as in many archaic cultures (among them that of the early Slavs), the mythopoetic world view was predominant and not only cultural texts but natural events, too, were interpreted in terms of sign systems. In the rational European tradition, following the scientific revolution of the seventeenth century and the Age of Enlightenment, it became necessary to separate these two sets of data. Yet even in modern societies the observer's attitude has remained the chief point of reference in finding—or measuring—information contained in a message. Thus recent semiotic research to be discussed in *Elementa* draws on psychological, neuropsychological and psychoanalytic studies (as was anticipated by Lev Vygotsky and Edward Sapir). The role of the pragmatic aspect is particularly important for the investigation of the potential influence of the signified facet of the sign on the signifier, as was suggested by Pavel Florensky (who as an example cited the miracle-working icon). From this point of view the sign ceases to be absolutely arbitrary: its relative arbitrariness depends on the mutual relations between its users and the levels of a given sign system. The number of such levels in different types of signs may provide another topic of the cross-cultural semiotic comparison to be reflected in *Elementa*. Thus, for a modern narratological theory it may become necessary to explore whether the distinction between the plot and the story-line (fabula) is valid for different cultural texts and contexts.

Natural language, literature, folklore, mythology, ritual, religious and philosophical concepts, scholarship, science and technology, visual and audiovisual arts, ornaments and dress codes may all be considered constituent parts of a cultural whole. Since every culture consists of a set of intersecting and coexisting texts, signs, sign systems and sign complexes, cultural typological studies are based on semiotic principles. The aim of current cultural semiotic comparison is to identify general, indeed universal features (e.g., broadly applicable schemes of rites, plots, motifs, archetypes such as the world tree, and other symbolic models of the universe) as well as those elements and rules of their combination that are characteristic of some particular area, community or individual. Any historic event or personality may be seen as part of a particular symbolic tradition. Their relation to their historical context will be among the most intriguing semiotic questions to be investigated on the pages of *Elementa*. It is our position that the reconstruction of the past should become the most important task for some branches of scholarship. From this point of view, semiotic diachronic studies—using exact methods like those of comparative linguistics and of Alexander Veselovsky's historical poetics—may be compared to those fields of historical, archaeological and psychological research of the past that has become so significant by the end of this century. The value of verbal and non-verbal symbols is connected with an even more intensive search for methods of forecasting and understanding the future. The evaluation of different methods of social engineering as opposed to other ways of regulating and controlling the development of society becomes an integral part of applied knowledge.

Continuing the tradition of their predecessors in Prague, St. Petersburg, Moscow and Tartu, the contributors to *Elementa* will discuss the general principles of the semiotic study of culture as applied, above all, to Slavic texts, symbolic systems and individuals. The common background of the earlier Baltic and Slavic (as well as original Indo-European) mythopoetic tradition is being reconstructed by combining linguistic, folkloristic and other semiotic methods. Thus, it has already become possible to recover not only words and their combinations but also some fragments of the mythological, epic and other folkloric texts which are instrumental in recreating the perception of the world as viewed by the ancient Slavs. Consequently, important comparisons of Slavic folk art have become possible. The early Slavic mythopoetic tradition reconstructed on the basis of linguistic, poetic and mythological data sheds new light on later historical transformations prompted both by internal factors and by cultural dialogical relations with other Eurasian traditions. For the medieval period, the evolution may be seen from the point of view of religious, cultural and other semiotic differences which led to the division of Slavdom into several zones oriented either toward Western (Catholic) Christianity (*Slavia romana*, where Latin remained the sacred language) or toward its Eastern variant (*Slavia orthodoxa*, with Church Slavonic as the sacred language). The coexistence of Eastern Christianity with paganism produced the phenomenon of "dual faith"—a particular type of religious syncretism—which is linked and comparable both to diglossia (the simultaneous use of two functionally different languages within one and the same community, e.g., Latin and Polish, or Church Slavonic and Russian) and heterography (the use of a writing system devised for one language to render the sounds of another one). The interconnection of these three phenomena seems particularly revealing for general cultural semiotic theory.

In studying modern Slavic sign systems, it is possible to find traces of the pre-Christian and early Christian epochs with other elements, the latter due to influences from, on the one hand, Western and Central European and, on the other, Eurasian traditions. In this way a synchronic analysis of contemporary Slavic sign systems may be informed by a diachronic commentary to their history. Primary data found in archives only recently made available will be investigated in this connection.

One of the general problems of recent semiotic study conducted by members of the Moscow–Tartu school has been the relationship between the creative abilities of man and the restrictions imposed on him by social institutions and systems. In the history of Slavic cultures the role of one individual (such as Pushkin or Tolstoy in Russian literature) is surprisingly significant. Some of the unique Slavic figures (say, Dostoevsky) had an enormous impact on the cultural development of the twentieth century as a whole. Another important aspect is the role played by a small group of great creative minds (such as the Polish romantic poets in exile during the nineteenth century) who may have laid the foundations for future traditions. By the same token, it is possible that some of the most gifted individuals (e.g., Cyprian Norwid, to continue the last example) were not understood and accepted by their own semiotic environment precisely because they transcended its limits. All of the avant-garde movement in the Slavic countries may, in fact, be seen in this light. This was anticipated by exceptional personalities of the previous century. Representatives of both the arts and the sciences in the first half of the twentieth century seem to have been interested primarily in syntactic (formal) relations and structures, paying less attention to issues of sense and meaning, that is, to semantic and pragmatic aspects, which instead have become more central to semiotic studies and artistic creation during the second half and toward the end of this century. The artistic avant-garde may be seen as part of the general cultural movement to which Russian, Czech, Polish and other Slavic precursors of modern structural and semiotic studies belonged. Their work—partly as yet unpublished and often not sufficiently known—will be studied with great attention. The emphasis throughout will be on the interconnections within one culture and the relations among various cultures.

Vyacheslav Ivanov

Los Angeles/Moscow

On The Etymology of Latin *Elementa*

VYACHESLAV IVANOV

University of California, Los Angeles; Moscow University

THE LATIN *element-um* (singular, plural #*-a*) is usually explained by the archaic Roman (possibly originally Etruscan) habit of dividing the alphabet into two parts for teaching purposes (Coogan 1974, 1990). The first of these began with A B C and gave rise to such names as "abecedary" or "alphabet." The other one had the sequence L M N as its initial group of letters. The exact numerical equality of the two parts of the alphabet can be supposed only for the Proto-Canaanite system (preserved in early Phoenician, Hebrew and Aramaic traditions) consisting of 22 letters; the second part of it began with the sequence L M N and ended with T (Naveh 1987: 180–181). To understand the last *-t-* in *elemen-t-um* one may compare it to the possible Etruscan name of the alphabet *abat* (Lejeune 1984, 1989) in which the first two letters *ab* are combined with the same *-t-*. The Etruscan intermediary role supposed also for the name of "letter" (Latin *littera*, Sandoz 1991) and in general for the introduction of writing in ancient Italy (Lejeune 1957; see Haarman 1990: 294–296) might be responsible also for the pronunciation of the names of the letters *el, em, en* < syllabic l, m, n (Thurneysen 1934) that are so different from the original West Semitic (and Proto-Canaanite). The later number of Latin letters (21) and their distribution inside the two groups was different from earlier Greek (23, Gamkrelidze 1989: 232; later 27) and Etruscan (26) (Flobert 1991: 525). But still it is known from Pompeian graffiti that the alphabet was taught by learning two parts of it that could be combined in direct and reverse order as in the inscription CIL, IV, 5499:

A X B V C T D S E R F Q G P H O I N K M L

Elementa, 1993
Vol. 1, pp. 1–5
Reprints available directly from the publisher
Photocopying permitted by licence only

The crucial role of the sequence L M N is seen from this fixation of the learning device.

Thus it seems that if the Etruscan or Latin *element-a* was reshaped later from an older name of the 'ivory tablet' (Puhvel 1993), it is true that the later popular etymology was rooted in the old structure of the Roman and Etruscan alphabet. As Jaan Puhvel writes in his letter to me of May 26, 1992, "already Diels (1899: 70–81) connected *ἐλέφας* with reference to Quintilian, Inst. orat. I.I. 26, '*non excludo autem id quod est inventum irritandae ad discendum infantiae gratia, eburneas etiam litterarum formas in lusum offerre.*' The idea that *elementa* were ivory letters for learning the alphabet was further developed by Vollgraff, 1949. At the most I might make the additional observation that Etruscan ivory writing tablets have the alphabet inscribed along the edges for learning purposes, such as the one from Marsiliana d'Albegna."

If according to this hypothesis *element-um* is based on a former name of the ivory tablet with letters (they might have been originally made of ivory as a luxury, but were later on replaced by cheaper metallic letters known in Rome: Vollgraff 1949: 113–114; Vernadskii 1981: 107; Ivanov 1988: 49; Carlo Ginzburg pointed out to me an interesting analogy in edible letters in Florence described by Bocaccio, see Luchi 1978: 627–628), the change of the phonemic shape was due to the function of the object designated by the word: *ἐλέφαντα* > Etruscan-Latin *elementa*. As Jaan Puhvel suggests, Etruscan (like Romance languages) had no equivalent "for the Greek and Latin nominative:accusative opposition generalizing the accusative: νάβλας: *naplan*, χρυσηΐδα: *Crisida*, κρηπῖδα > Lat. *crēpida*." The difficult problem of the change *b(ph)/m* in this word (see Diels 1899: 85–86; Vollgraff 1949: 113) may be traced back to its Oriental origins: the Hittite *laḫp-ant*, according to E. Laroche a cognate of Greek *ἐλέφας* (Ivanov 1977, 1984) and Latin *elephantum* 'ivory,' has a possible variant *laḫma-*. But according to Jaan Puhvel "it would suffice to lump the φ : m variation with the general Etruscan tendency to confound stops and nasals (see *Melerpanta* <Βελλοροφόντος, *Marmis* < Μάρπησσα, Lat. *forma*<μορφή, *formica* vs. μύρμηξ etc.)."

The later semantic development of the Latin word *element-um* (*-a*) (see Diels 1899; Lagercrantz 1911) and its derivatives in modern European languages was connected with its use to render the Greek στοιχεῖον (-α) in the meaning of the main parts of the universe (in Plato or later tradition) or of the human organism (in Galen and his followers). The latter has proved to be particularly important for European psychology and poetics. The former meaning has remained valid for the whole scientific tradition from Euclid (the title of whose work was translated into Latin as *Elementa*) to Mendeleev (whose study shows the importance of the term for the Russian scientific language of the nineteenth century). It can be supposed that Mendeleev was developing, in the idea of a *tablitsa elementov* (periodical table and system of elements), the terminological tradition of early chemical and alchemical treatises: thus *elementa* are combined with the word *tabula* "table" in a Latin alchemical treatise by Robbert Fludd in the beginning of the seventeenth century (Fludd 1629: 134–142). But in both cases the term *element-um* and its modern European continuations (for example, Russian *element*, which designated before Pushkin—but not later—the main force of nature in the poetry of the eighteenth century: *vsekh stikhii* 'all the elements' in Lomonosov's ode of 1752; Polish *element*, having the meaning of στοιχεῖον in Old Polish texts, etc.) are representative of the way of thinking that has tried to analyze the world and the human psyche down into its ultimate constituents regarded as discrete units. In this way the term seems to be appropriate also in modern studies of the semiotics of culture, particularly of cultures based on the notion of the alphabet (Ivanov 1990). In an early eighteenth century translation of Plato's dialogues the Greek στοιχεῖα 'elements' was systematically rendered by the Russian word of the same Greek origin *stikhii*, plural of *stikhiia*. The Russian form comes from the Old Church Slavonic one with -y- rendering Greek -y-/-oi- (Diels 1963: 29, §8, fn. 8; the Old Church Slavonic form is attested in Supr. 475, 12; Euch. 4a, 1; Chil. 5, 9). It appears, for instance, in the final part of "Theaetetus" where the constitution of the syllable is discussed (Platon 1783: 85–86, 88–91). Later, especially in the Russian poetic language, *stikhiia* (pl. *-i*) 'element(s)' may still refer to the natural forces (see in Pasternak's *Themes and Variations*, "stikhiia svobodnoi stikhii/ s svobodnoi stikhiei stikha"—"The element of the free elements with the free element of verse"). Russian *stikhiia* is common in expressions like *v svoei stikhii*–"in his/her element" instead of the older *v svoem elemente*. But in the meaning of a part of a certain set usually the Russian term *element* (of the Latin and Western European provenience) is used. Thus the double (Greek and Latin) character of the Russian abstract vocabulary is clearly seen in this branching of possible translations of *elementa*. Another (Church) Slavonic equivalent, the archaic *nachal-a* might be found in some texts (where, as in the translation of Newton's work, it renders the Latin *principia* synonymous to *elementa*).

According to researchers who have shown the role of *element-um* to be the equivalent of Greek *στοιχεῖον* (Lagercrantz 1911; Vollgraff 1949) this term was particularly important for the atomistic view of the world. Since the structure of the universe was seen through the metaphor of the separate letters being combined with each other, one may view this use of *elementa* as one of the early approaches to semiotic description. To describe the discrete units of which the utterance or a fragment of the world was built meant to attempt introducing what is now called an "emic" level of description. Although it may be supposed that not all systems or texts described by modern cultural semiotics really consist of discrete units, still our methods to study them continue the logical way of reasoning discovered and developed in the Western Semitic and the later Greek, Etruscan and Roman traditions.

Acknowledgment

The author wishes to express his gratitude to J. Puhvel for his most valuable remarks and suggestions incorporated into the text and to C. Ginzburg for his comments.

Works Cited

Coogan, M. D. 1974. "Alphabets and Elements." *Bulletin of the American Schools for Oriental Research*, no. 216: 61–63.

______. 1990. "*'lp- "To be an Abecedarian"." *Journal of the American Oriental Society*, vol. 110, no. 2: 322–323.

Diels, N. 1899. *Elementum. Eine Vorarbeit zum griechischen und lateinischen Thesaurus*. Leipzig: B. G. Teubner.

Diels, P. 1963. *Altkirchenslavische Grammatik*. 2 Aufl. Heidelberg: C. Winter Universitätsverlag.

Flobert, P. 1991. *L'apport des inscriptions archaïques à notre connaisance du latin prélittéraire*, Latomus, t. 50., fasc. 3, 521–543.

Fludd, R. 1629. *Medicina Catolica, seu Mysticum artis medicandi sacrarium*, vol. 1, sec. 1, part 1, Liber primus. Frankfurt.

Gamkrelidze, Th. V. 1989. *Alphabetic Writing and the Old Georgian Script. A Typology and Provenience of Alphabetic Writing Systems*. Tbilisi: Publishing House of the Tbilisi State University (in Georgian with a long Russian summary).

Haarmann, H. 1990. *Universalgeschichte der Schrift*. Frankfurt/New York: Campus Verlag.

Ivanov, V. V. 1977. "Nazvaniia slona v iazykakh Evrazii. 3. Khet. *lahpa*- 'slonovaia kost' ' i grech. ἐλέφας" (Names of 'elephant' in languages of Eurasia. 3. Hittite *lahpa* 'ivory' and Greek *ἐλέφας)*. In: *Etimologiia* 1975. Moscow: Nauka, 158–161.

______. 1984. "O proiskhozhdenii dr.-grech. ἐλέφας 'slonovaia kost', slon'" (On the Origin of Greek ἐλέφας 'ivory, elephant'). In: *Etnogenez narodov Balkan i Severnogo Prichernomor'ia. Lingvistika, istoriia, arkheologiia*. Moscow: Nauka.

______. 1988. "V. I. Vernadskii i istoriia drevnosti." (V. I. Vernadskii and the history of the Old World). *Vestnik drevnei istorii*, no. 4: 38–65.

______. 1990. "Alfavit" (Alphabet). In: *Lingvisticheskii entsiklopedicheskii slovar*. Moscow: Sovetskaia entsiklopediia, 29–30.

Lagercrantz, O. 1911. *Elementum: ein lexikologische Studie (Skrifter utgivna av K. Humanistiska vetensskapssamfundet, 11: 1)*. Uppsala: Akademiska Bokhandeln.

Lejeune, M. 1957. "Sur les adaptations de l'alphabet étrusque aux langues indo-européennes d'Italie." *Revue des Études Latines*, vol. 35: 85–105.

_____. 1984. "Notes de linguistique italique. XXXIV, Un nom étrusque de l'alphabet?" *Revue des Études Latines*, vol. 59: 77–79.

_____. 1989. "Le jeu des abécédaites dans la transmission de l'alphabet." In: *Secondo Congresso internazionale etrusco*, vol. 3, Roma, 1285–1291.

Lucchi, P. 1978. "La santacroce, il salterio e il babuino. Libri per imparare a leggere nel primo secolo della stampa." *Quaderni Storici, 38. Alfabetismo e cultura scritta*: 592–629.

Naveh, J. 1987. *Early History of the Alphabet. An Introduction to West Semitic Epigraphy and Paleography.* 2nd ed. Jerusalem: The Hebrew University, The Magnes Press.

Platon. 1783. *Tvorenii Velemudrogo Platona chasti vtoryia pervaia polovina, prelozhennaia s grecheskogo iazyka na rossiiskii sviaschennikom i Imperatorskoi Rossiiskoi Akademii chlenom Ioannom Sidorovskim i Sekretarem Matfiem Pakhomovym, nakhodiaschimsia pri obshchestve blagorodnykh devits.* V Sankt Peterburge: izhdiveniiem Imperatorskoi Akademii Nauk.

Puhvel, J. 1993. "The Tocharian word for 'elephant'." *Historische Sprachforschung* (forthcoming).

Sandoz, C. 1991. "Le nom de la 'lettre' et les origines de l'écriture à Rome." *Museum Helveticum*, fasc. 4, vol. 48: 216–219.

Thurneysen, R. 1934. "Elementa." *Zeitschrift für vergleichende Sprachforschung auf dem Gebiete der indogermanischen Sprachen*, vol. 61: 255–256.

Vernadskii, V. I. 1981. *Izbrannye trudy po istorii nauki (Selected Works on the History of Science).* Moscow: Nauka.

Vollgraff, W. 1949. "Elementum." *Mnemosyne*, vol. 4, no. 2: 89–115.

GENERAL PROBLEMS OF SIGN AND CULTURE

Language, Reality, Culture and Semiotic Modeling

HENRIK BIRNBAUM

University of California, Los Angeles

THE LINGUISTIC, literary, and broadly semiotic studies produced by scholars associated with the Moscow-Tartu school of semiotics—and, more recently and specifically, of semiotics of culture—constitute, taken together, one of the most remarkable accomplishments in the field of humanistic learning during the second half of the twentieth century. This school traces its roots from the dual, yet intimately interrelated, tradition of linguistic structuralism and literary formalism which was suspended, if not outright prohibited, during the heyday of Stalinism and also repressed in the subsequent years of the Khrushchev era (for some concrete instances, see Ivanov 1991). It was, moreover, influenced and enriched by the unorthodox—in part even "subversive," albeit thinly disguised—thought of M. Bakhtin and his circle as well as by insights gained from advances in cybernetics (the cognitive sciences and artificial intelligence) and neurology. Still, because the Moscow-Tartu school has retained a dynamic, ever developing and integrative conceptual framework, it never turned into a rigid, once-and-for-all established doctrine inaccessible to new ideas and creative exploration.

The project embarked upon by Viacheslav Ivanov, one of the most prominent members of the Moscow-Tartu school, indeed, along with Iu. Lotman, its chief theoretician, to extend and continue the discussion of the ideas and notions of the school, and their promulgation, on a truly international, worldwide scale by launching *Elementa. Journal of Slavic Studies and Comparative Cultural Semiotics* is to be hailed as a significant and highly welcome enterprise. It is also most timely given the fact that the twin home of the original school, at the Institute of Slavic and Balkan Studies of the Soviet (as

Elementa, 1993
Vol. 1, pp. 9–22
Reprints available directly from the publisher

of late, Russian) Academy of Sciences in Moscow and at Tartu University (with its nearby summer school at Kääriku), is now split, literally, between two independent countries—Russia and Estonia. Also, many of the school's former members and followers have moved to various other parts of the world.

Having previously discussed some of the tenets propounded by the Moscow-Tartu scholars which bear on major theoretical issues of semiotics, I greatly appreciate the opportunity to summarize some of my pertinent views and to resume my assessment of their findings and achievements in the pages of this journal.

Let me begin with the notion of linguistic creativity. In an earlier study (Birnbaum 1982), I conceived of what could be thus labeled as two essentially separate, qualitatively different processes which, however, from another perspective could also be considered two consecutive, interdependent stages (or phases) of one and the same phenomenon. If we adopt the broadly defined notion of deep structure where semantic-syntactic categories—i.e., meaningful-functional elements markedly more abstract than the items of any particular lexicon—obtain and relate to one other (without as yet being ordered sequentially, along the syntagmatic time-space axis, such ordering taking shape at the several levels of surface representation—from the syntactic to the phonetic and graphic; Birnbaum 1975, 1977), the first stage of linguistic creativity can be said to be generative in the now conventional, though here loosely used sense. Thus, as I have previously suggested (Birnbaum 1982: 77), "the notion of creativity with reference to language has . . . been invoked . . . to denote the capability of man—innate in the opinion of some—to select (from a code internalized and stored in the human brain), to fashion, and to concatenate meaning-differentiating elements (alias phonemes) and minimal meaningful entities (alias morphemes) in sound sequences forming discrete speech segments (usually words) and conveying a specific, if sometimes ambiguous, message. In this sense, then, linguistic creativity is roughly equivalent to man's ability to generate well-formed sentences." And I went on to point out that "the generative approach to language as a functioning system is by no means . . . dating only from 1957, the year of Noam Chomsky's *Syntactic Structures*," indicating that W. von Humboldt's view of language as 'energeia' as well as 'ergon' and R. Jakobson's 1948 reanalysis of the Russian conjugation (positing but one basic verb stem on which specific rules operate in order to generate all actually existing verb forms) implied the same, essentially generative conception; these two telling examples of such an approach are by no means the only ones, needless to say. Here it can be noted that the great Russian polymath P. Florenskii, influenced by the linguistic thought of A. A. Potebnia, in an original way developed von Humboldt's notion of language as 'energeia' and also his doctrine of 'inner form.' See notably his study "Stroenie slova," conceived as a chapter in a larger work to be titled *U vodorazdelov mysli*; for some further discussion, see Ivanov 1988 and Cassedy 1991 (esp. 543–547). Compare further Florenskii 1988 (esp. 92 and 98), another posthumously published chapter of the planned, larger philosophical treatise. In my earlier treatment I, moreover, suggested that F. Jameson's facile assessment to the effect that "the originality of Chomsky's transformational grammar seems to derive from a reversal of the Saussurean model, a kind of negation of the negation in which the linguistic mechanisms are relocated back in the *parole* or individual act of speech" (Jameson 1974: 26, fn. 22) was not sufficiently comprehensive and therefore also not entirely accurate. In this context reference can also be made to the notorious speech act controversy between N. Chomsky and J. Searle.

The other kind of linguistic creativity, presupposing the first one, implies, of course, the rare ability and activity of producing genuine works of verbal art—in other words, artistic creativity, properly speaking, as well as various combinations of perception and aesthetic experience. As I stated with reference to this (Birnbaum 1982: 77), "though aesthetic values are not readily susceptible to exact measurement and objective assessments, works of art can nonetheless be subject to scientific analysis revealing and establishing at least a number of criteria and characteristics inducing pleasure and enjoyment: the Kantian 'interesseloses Wohlgefallen.' Thus, the study of poetic language can, among other things, be conceived as part of semiotics, here viewed in its aesthetic function." I then went on briefly to analyze Iu. Lotman's notion of the poetic text, focusing in particular on his attempt at defining the—derived—aesthetic nature of (artistic) prose which "simultaneously unites the idea of high art and non-poetry" and behind which "stands the esthetic of 'real life' with its conviction that the source of poetry is reality" (Lotman 1976: 24). Incidentally, a very similar conception of artistic prose, its role and function, also underlies B. Pasternak's later writings, notably *Doctor Zhivago*, as I have argued elsewhere (Birnbaum 1980, 1989, 1991a).

Briefly touching upon Lotman's and his fellow semioticians' notion of primary and secondary modeling systems, external versus internal communication (of which the former serves to transmit messages whereas the latter aims at doing the same with codes), and Jakobson's model of verbal communication, I subsequently (Birnbaum 1982: 82–84) ventured to take a somewhat closer look at an original approach to language and music by

Leonard Bernstein. Specifically, the famous American composer and conductor, intuitively rather than strictly scientifically, compared how linguistic texts, including poetic texts, are produced and what generative mechanisms are at work in musical composition. In doing so, he drew heavily (and too one-sidedly, in my opinion) on Chomsky's 1972 essay *Language and mind*. While I saw some merit in the parallelism observed, I also could not but express some reservations, pointing out that "it remains arguable . . . whether the transformational approach of linguistics adopted by Bernstein to explicate the creative process of composing music (and, in some measure, also the re- and co-creative activity of . . . listening to music with joy) is indeed the only viable or even an adequate one." For even if, for argument's sake, we go along with Bernstein's tentative supposition regarding the parallelism between linguistic and musical creativity, it should be noted that, whereas in music the 'transformational leap' from 'musical prose' (at the deep structure level) to a "totally metaphoric language" is always present, this process can, but must not occur in natural language, only this further 'transformational leap' yielding poetry, in verse or prose (as opposed to ordinary speech, this latter resulting from the kind of linguistic creativity referred to above). For a strictly generative approach to musical composition, see also Lerdahl and Jackendorff 1983 (esp. 1–11, 286–290 and 317–332).

In this context I subsequently turned to Jakobson's well-known definition of poetic language and to his crucial distinction—on a sliding scale, surely—between metaphor and metonymy (ibid.: 84–87) and commented in conclusion on V. Shklovskii's notion of 'defamiliarization' (*ostranenie*) as the chief artistic device, in prose, at any rate. Here, I also mentioned the second artistic technique uncovered by Shklovskii—that of 'retardation' (*zaderzhanie*) or 'impeded form' (*zatrudnennaia forma*)—and pointed out that the Russian critic, contrary to claims based on careless reading of the opening section of his famous essay "Art as device" (*Iskusstvo, kak priem*, first published in 1917), "never tried to do away with the notion of image in verbal art: he merely distinguishes between 'prose images' (the proper domain of 'thinking in images') and 'poetic images' (roughly the equivalent of metaphors)." I also remarked on the affinity found between Shklovskii's 'defamiliarization of image' and Jakobson's view, consistent throughout his scholarly career, of the poetic function of language, going back to his 1921 article "On artistic realism" (first published in Czech).

On a different occasion (Birnbaum 1985), I suggested that in addition to 'defamiliarization' its very opposite, what thus could be termed 'familiarization' (as a semiotic-aesthetic category), also can serve the primary purpose of increasing the artistic effect by focusing on something previously unfamiliar, threatening, inescapable and irreversible, while signaling the end of a familiar situation and physical presence taken for granted—in short, non-existence in its absolute, ontological sense. This device, I argued, applies equally to prose and poetry (the latter here taken in the narrow, formal sense). To illustrate its use, I referred to the Finish-Swedish poet Edith Södergran's two last poems, "Landet som icke är" (The land that is not) and "Ankomst till Hades" (Arrival in Hades), and provided a somewhat more detailed analysis along these lines of A. Akhmatova's poem "K smerti" (To Death) and P. Celan's "Todesfuge" (Death Fugue).

Rather than elaborating further on the poetic devices of 'defamiliarization' and 'familiarization,' I would like now to comment briefly on Jakobson's oft-quoted and much-discussed dictum, stating that "the poetic function projects the principle of equivalence from the axis of selection into [sic] the axis of combination" (Jakobson 1960: 358), in passing also referred to by me (Birnbaum 1982: 84) and more recently reevaluated at length by H. Andersen (1991). It is Andersen's contention that what he calls Jakobson's Projection Principle (and the aesthetic Projection Thesis derived from it) is, for one thing, by no means only a narrowly linguistic principle since it is manifested in a variety of other types of semiotic behavior as well, and that, second, and as a consequence of the principle's much wider application than to language only, Jakobson's Projection Thesis—i.e., the claim that the Projection Principle produces aesthetic value in poetically shaped messages—by itself is not really valid. Also, the verb 'to project,' though having some merit, is considered by Andersen much too vague, not to say vacuous; compare, in this context, a similar terminological objection raised by me concerning Jakobson's notion *Gesamtbedeutung*, or 'general meaning' (citing also criticism by S. D. Katsnel'son and A. Wierzbicka; Birnbaum 1986). However, Andersen—perhaps in the spirit of the central notion of complementarity embraced by Niels Bohr (with whom Jakobson at one point closely collaborated, incidentally)—finds that Jakobson's theory of 'introversive semiosis,' though flawed by the misconception of adding a fourth sign type to Peirce's well-known trichotomy (of icon, index and symbol), does in fact capture an essential facet of all art, namely, the automorphic structure which is made up of reciprocal iconic indexes (rather than 'signs of imputed similarity,' as Jakobson had conceived them). For it is precisely the reciprocity of iconic indexes involved that lends a work of art its character of 'introversion,' or, to be precise, automorphism. In other words, contrary to purely mimetic icons which depict or reflect the reality represented, and to projective diagrams which represent relations in the code of which a particular work is an instantiation, automorphic signs (iconic indexes) represent one another and are themselves represented by one another. So perceived, the constitutive parts of an automorphic structure, which are conducive to being assigned aesthetic interpretants, form a single sign whose aesthetic sign-object triggers or releases an experience (or, more strongly, a sensation) of joy, pleasure and satisfaction. While conceived for two different purposes—the Projection Thesis to account for verbal art and the theory of 'introversive semiosis' for non-verbal art forms—Jakobson's two aesthetic theories are indeed complementary, Andersen suggests, also in the sense that the former focuses on the encoding, the latter on the decoding process. In the second instance this applies, however, without taking into consideration the paradigmatics from which selection was made for the purpose of combination, i.e., syntagmatic concatenation.

Perhaps the above brief account of Andersen's line of reasoning does not do full justice to his more detailed and precisely phrased argument pertaining to Jakobson's aesthetic theories; hopefully, though, it at least may convey its gist. For my part, I see considerable merit, indeed persuasive power, in Andersen's qualified critique of Jakobson's relevant thought. In particular, it should be noted that Andersen recognizes the significance and insightfulness of Jakobson's reasoning for semiotic behavior and orderly rule-governed activity at large. But it does not apply specifically to aesthetic semiosis where Jakobson's Projection Principle is found wanting as it is all too generally worded and therefore takes in aesthetically significant along with insignificant projective diagrams.

In connection with Jakobson's semiotics, I further submit that he has, generally speaking, gone too far when, inspired by Peircean semiotics, he overemphasized the role of iconicity at the expense of the Saussurean notion of the arbitrariness, in most instances, of the linguistic—or, for that matter, any conventional, imputed or habitual—sign. For even if we accept Peirce's fundamental trichotomy of all signs (among them—with some qualification, see below—also the linguistic sign) as categorizable into icons, indexes and symbols, I feel that Jakobson, always deeply concerned and fascinated with the relationship of sound and meaning, particularly in his influential semiotic paper of 1965 (Jakobson 1966), gave the notion of iconicity and diagrammatization in language an unduly broad range and application. This is not to say that I am not duly impressed, as are so many others, by his brilliant reasoning, set forth with great ease, in his "quest for the essence of language." Yet I would argue that some of the examples adduced by him in support of his iconicity claim, while acknowledging the "sliding," gradated division of Peirce's threefold categorization of signs, are not entirely compelling. This applies, among other things, to his interpretation of the longer plural forms (of nouns as well as verbs) in many languages as supposedly indicative of the larger number of individuals referred to, where corresponding earlier attested (or reconstructed) forms frequently would not bear out such a semiotic relationship and sometimes might even be construed as demonstrating the opposite. For the claimed principle would, obviously, have to apply equally to any phase of linguistic evolution, assuming that this is not contradicted by the teleological view of the development of language. As is well known, Jakobson explicitly held precisely such a view which was bolstered with additional arguments by M. Shapiro; for this semiotically informed teleological interpretation of linguistic change, implying goal-directedness and 'final causation,' see Jakobson 1963 and 1975, and more recently, using a fuller Peircean argumentation, Shapiro 1991 (esp. 22–91). For my part, I continue to have some reservations about the assumption of an unqualified application of the teleological principle to language evolution. Because I believe that the two fundamental factors affecting all linguistic change, namely, greatest possible economizing with the means of expression available (sound) and a striving for optimal communication of content (meaning), while mostly keeping a precarious balance, are not necessarily always directed toward, or subordinated to, one single goal. Also, if such iconic relationship between sign and object is to be considered a genuine language universal, it would have to be observable everywhere in language regardless of any specific evolutionary stage of a particular language. Other examples, beside the plural forms, where the semiotic interpretation is not as obvious and unequivocal as claimed in this instance, could easily be adduced, but space considerations preclude such elaboration here. For further illustration with Russian data, see, e.g., Shapiro 1969. In this context, it is worth pointing out that contemporary, highly sophisticated literary scholarship continues to consider the linguistic sign (and other symbols) as essentially—or, at any rate, in most instances—arbitrary, that is, as mostly based on convention in the Saussurean sense, or, in Peirce's understanding, as owing to 'symbolic law' (for details, see below). Compare in this vein, for example, the thrust of the recent book by M. Krieger on 'ekphrasis' (Krieger 1992). For some further observations on the semiotics of poetic language, also from a poststructuralist point of view, note further my remarks in Birnbaum 1988 and 1990a.

Turning now to some of the basic tenets of the Moscow-Tartu school, I have already had an opportunity to discuss two particular points which, in my view, need further elucidation and where, notwithstanding my genuine appreciation of the school's overall achievements, I have found myself at variance with what appeared as its theoretical stance, at least at an earlier stage of its development. However, in this connection, I would like, once again, to point to the Moscow-Tartu scholars' dynamic, ever-developing conceptual framework open to fresh ideas and creative exploration. This was recently confirmed by V. Ivanov's statement (1991: 35) that "the most interesting aspects of our first semiotic publications and conferences in Moscow and Tartu were not so much our theoretical views, which have changed considerably since then, but the technique and materials used for the study of all these and some other systems."

My first point concerns some aspects of what the Russian semioticians have referred to as 'primary' and 'secondary modeling systems' (or by equivalent terms). In a recent paper on the subject (Birnbaum 1990b), while finding fully justified the distinction between primary and secondary semiotic modeling systems—or if, with T. A. Sebeok, we posit a yet more immediate (primitive) level of modeling (or representing) reality by means of non-verbal, zoo-semiotic signs, even three qualitatively different levels of modeling reality (or the 'world')—I indicated that I cannot accept the idea that only natural language functions as the primary (or, in Sebeok's terms, secondary) modeling system underlying all other secondary (or tertiary) systems involved in creating—or generating—cultural values of the most various kinds. I here elaborated and echoed my earlier reservations, voiced in an essay contributed in honor of V. Erlich (Birnbaum 1985) where, among other things, I had argued (p. 156) that "it is not sufficient to regard only natural language as the primary modeling system representing reality. For in addition to ordinary language there are obviously other, equally primary means to represent the real world and its extension into the realm of ideas and emotions." I went on by making reference to photography in all its forms, from snapshots and stills to cinema and TV, which, like language, could be used both as a purely communicative or investigative, mere reality-depicting technique (see, e.g., holography or photographic microscopy used in biological and medical research, on the one hand, and, on the other, photographically recorded telescopy applied in astronomy and space science) or as an art form, or even, arguably, several art forms. And, while noting the great interest that S. Eisenstein took in the aesthetic and semiotic theories of the Russian formalists, I also mentioned the difficulty of always clearly distinguishing between the artistic and non-artistic in film as in fiction. Here, it could be argued, with Lotman, that it is precisely the blending of pure reality-reproduction with artistically shaped language or some artfully focused picture sequence, respectively, that produces the—modern—aesthetic impact in the first place. By the same token, there are of course a number of areas of human endeavor—in the sciences, in particular (see, e.g., cartography, topography, geodesy, geometry, astronomy, or anatomy)—where one normally does not go beyond attempts at primary modeling of segments of reality, though here, too, secondary uses, for the purpose of secondary (artistic) modeling are conceivable. Compare as examples the anatomic or perspective measurements of masters in the visual arts—say, Leonardo or Dürer—preliminary to their creating graphic or otherwise pictorial works of art. By contrast, the expressionist "starry nights" by van Gogh and Munch are obviously based on direct observation by the naked eye only, i.e., without any telescopically based measurements or astronomic calculations. I have further elaborated on my disagreement with the Moscow-Tartu scholars' perceived position concerning the primacy of natural language as a modeling system in another paper where the focus was on the semiotics of poetic language (Birnbaum 1990a).

In this context I was pleased to note that V. Ivanov in his recent autobiographical sketch remarked that, while "the formation of the Moscow-Tartu school was made possible by our common acceptance of the linguistic model, according to which semiotics might develop methods close to those of the modern science of language, many years later all of us came to understand that there are several models, complementary in respect to each other." And he continues by stating that "to me as to Lotman, cinema constituted an interesting example of a sign system which could not be understood only in terms of language" (Ivanov 1991: 36). I therefore now feel that our potential theoretical disagreement in this respect has been resolved, particularly as Ivanov in a recent presentation at UCLA (10 April, 1992) of his and his Moscow-Tartu colleagues' semiotic thought confirmed his current view that natural language, though one of the most crucial and common primary modeling systems, cannot claim monopoly in this capacity.

Strictly speaking, though, to consider natural language a primary modeling system implies a gross oversimplification, of course. For in actual fact it is the human brain that simulates reality not unlike a supercomputer and only subsequently generates the model thus created in order to produce, as the output of a complex process, chains of speech sounds. Also, generally, any computer simulation (and monitoring, say, of the flow of ground traffic on the freeway network of a large city or the air traffic into and out of a metropolitan airport) amounts to a primary modeling of a segment of reality (in these instances, largely in permanent motion). Or, to return to the human brain, if we are to accept G. M. Edelman's alternative view (referred to as "neural Darwinism"), information conveyed by our sensory organs, rather than being digitally processed, sets off biochemical reactions which strengthen the links between the neurons which fine-tune the circuitry we use to interpret the world; see Edelman 1992. Consequently, I would here merely reiterate my own view expressed previously (Birnbaum 1990b: 61–62) that "underlying all genuine human creativity, or in other words, the creation of cultural values of artistic and scientific imagination, are sets of primary systems, modeling—or selectively reproducing or shaping—reality, the real and our inner world, in accordance with man's innate unique sensory apparatus and the faculties of his perceptional equipment (Jacob 1982: 55). No doubt, ordinary language ranks very high, and perhaps highest, among these primary modeling systems (or 'languages' broadly conceived), but it is by no means the only system of signs encoded or decoded." And I added that any previous assignment to natural language of a unique role as primary modeling system is, presumably, less to be attributed to the realization that language is the only qualitative capability setting man apart from the rest of the animal kingdom—an assertion, which, however, I would now qualify at least to the effect that language may be the only *manifest* distinctive trait of humankind—than it is perhaps the result of the acknowledged fact that structural linguistics has in recent times been paradigmatic for other branches of the science of man. As an example I quoted the impact of Jakobsonian linguistics on cultural anthropology as represented by C. Lévi-Strauss. Also, it should not be ignored in this context that the Moscow group of scholars which subsequently formed part of the overall Moscow-Tartu school was in its core almost entirely made up of linguists. It is another matter that with the arrival of poststructuralism and its refocusing not on underlying invariant structures and patterns of human behavior but rather on the anomalies, decentered fringe phenomena, and the gaps or "cracks" in any perceived or posited structure, the attention of humanistic scholarship has shifted as well. While, when verging on outright scientific nihilism (as in some extreme forms of deconstruction), this more recent development may raise justified doubts, there is, needless to say, also a positive side to it: some of the truths hitherto taken for granted in some quarters may upon closer scrutiny turn out to be no truths at all—or, at any rate, in need of serious reconsideration. When it comes to language, and particularly to linguistic semantics, this way of thinking is, moreover, not entirely new; not only is it close to the linguistic-philosophical thought of the later Wittgenstein (as propounded notably in his *Philosophical Investigations*) but also to the contextual conception of language sketched by P. Florenskii; see Ivanov 1988; Cassedy 1991: 543–547, esp. 547.

My other concern as regards Moscow-Tartu semiotics relates to the very notion of the linguistic sign, or the sign in general. In a fairly recent monograph on the concept of the sign in Soviet semiotics, P. Grzybek (1989) pointed out that Jakobson, by attempting to adopt and integrate with his own Peirce's semiotic conception, in fact distorted the American philosopher's fundamental idea of the sign—or more precisely, the sign-object relation—as occurring in three different, yet usually somehow combined forms (thus shading into one another, as it were), namely, as icon, index and symbol, respectively. By retaining Saussure's notion of the dyadic nature, or two sides, of the linguistic sign and trying to accommodate to it Peirce's triadic interpretation, so Grzybek's argument goes, Jakobson, who otherwise had rejected or, at any rate, questioned several of Saussure's well-known dichotomies, did in fact misrepresent the essence of Peirce's thought and, in so doing, also affected negatively—in this one respect—the Moscow-Tartu semioticians whose own interpretation of Peirce was largely determined by Jakobson; see further Burg 1990.

In this context it should be mentioned that the notion of the dyadic nature of the linguistic sign was (as Ivanov informs us) prefigured by Florenskii as early as 1904, when he distinguished between its 'symbolizing' and 'symbolized' nature; for some further discussion of Florenskii's linguistic investigations, see Ivanov 1988. In my review of Grzybek's book (Birnbaum 1991b), I tended to agree with the German scholar's criticism. Note in this connection also Andersen's highly critical remarks concerning Jakobson's "attempt to analyse Peirce's concepts of icon, index and symbol into bundles of binary features" (Andersen 1991: 305). As Andersen points out, this effort produced "a misconstrual of Peirce's trichotomy." Essentially, I continue to share this view; however, one mitigating circumstance should be noted here. Saussure's (and, as we now know, Florenskii's) definition pertained specifically to the linguistic sign only, whereas Peirce's global sign theory was designed to embrace all signs used by man; see the notion of the 'semiosphere' in Lotman 1990: 121–214. Moreover, while Peirce's trichotomy of icon, index and symbol has been the most frequently quoted and by and large also the best understood—in no small measure, it should be pointed out, thanks to Jakobson (esp. 1966)—this represents only one of the several dimensions in Peirce's overall trichotomic scheme of representamens, or signs, subsumable under the headings of Firstness, Secondness and Thirdness, where a sign in relation to itself is termed 'qualisign' (or tone), 'sinsign' (token), and 'legisign' (type), respectively, whereas the more familiar Peircean terminology, strictly speaking, applies only to the sign in relation to its object, and yet another trichotomy—that of 'rheme,' 'dicent,' and 'argument'—pertains to the sign in relation to its interpretant.

Moreover, under Peirce's Firstness the sign is a possibility, under his Secondness an actuality, and only under Thirdness (which thus also subsumes 'symbol') the sign functions as a 'rule' or 'law.' Now it is my contention that linguistic signs (as opposed to other potential signs) are predominantly symbols functioning primarily as rules or laws, i.e., as part and parcel of a convention shared by a particular speech community. This seems borne out by Peirce's own threefold characterization of the category of Thirdness as implying mediation, rule (or law), and growth. For further discussion of the Peircean system of notions, see Shapiro 1983: 25–72. It follows, I would suggest, that Jakobson's misreading of Peirce was not so much a result of his effort to redefine his triadic semiotic model in binary (or dyadic) terms, as it was the undue weight he attached to Peirce's concept of sign as icon, given that Jakobson applied the American semiotician's general sign theory to the more restricted semiotics of language. I now feel that the disproportional role ascribed by Jakobson (and some of his followers, including his critics) to iconicity as a semiotic function of language derives from this misplaced emphasis. It is another matter that Saussure's (and Florenskii's) dyadic notion of the linguistic sign could be challenged from another point of view. What I have in mind is that of the logicians—from G. Frege through W. Quine—who, in addition to the signified, or 'denotation' ('meaning,' Frege's *Sinn*), also posited 'reference,' or 'connotation' (Frege's *Bedeutung*, or *das Gemeinte* as opposed to *das Bezeichnete* in E. Koschmieder's "noetic" terminology; see Koschmieder 1965, esp. 19–24, 26, 56–106).

In closing, I would suggest, as I have done before (Birnbaum 1990b: 53–55), that pertinent research has as yet insufficiently tapped the goldmine of semiotic thought found in E. Cassirer's magistral opus, *The Philosophy of Symbolic Forms*, which in more than one sense is very much in line with the Moscow-Tartu semioticians' subsequently developed ideas about primary and secondary modeling systems, without a direct influence being ascertainable, however.

Works Cited

Andersen, H. 1984. "Language structure and semiotic processes." *Arbejdspapirer* 3. Copenhagen: Institut for lingvistik, Københavns Universitet.

_____. 1991. "On the projection of equivalence relations into syntagms." In: *New Vistas in Grammar: Invariance and Variation*, Eds. L. R. Waugh and S. Rudy. Amsterdam and Philadelphia: John Benjamins (*Current Issues in Linguistic Theory* 49), 287–311.

Birnbaum, H. 1975. "How deep is deep structure?" In: *Proceedings of the Eleventh International Congress of Linguists*, Ed. L. Heilmann, II. Bologna: Il Mulino, 459–475.

_____. 1977. "Toward a stratified view of deep structure." In: *Linguistics at the Crossroads*, Eds. A. Makkai, V. B. Makkai, and L. Heilmann. Lake Bluff, IL/Padova: Jupiter Press/ Liviana Editrice, 104–119.

_____. 1980. "On the poetry of prose: land- and cityscape 'defamiliarized' in *Doctor Zhivago*." In: *Fiction and Drama in Eastern and Southeastern Europe: Evolution and Experiment in the Postwar Period*, Eds. H. Birnbaum and T. Eekman. Columbus, OH: Slavica, 27–60.

_____. 1982. "On linguistic creativity." *International Journal of Slavic Linguistics and Poetics* 25/26 (Festschrift E. Stankiewicz): 77–89.

_____. 1985. "Familiarization and its semiotic matrix." In: *Russian Formalism: A Retrospective Glance* (Festschrift V. Erlich), Eds. R. L. Jackson and S. Rudy. New Haven: Yale Center for International and Area Studies, 148–161.

_____. 1986. "*Gesamtbedeutung*—a reality of language or a linguistic construct?" *Scando-Slavica* 32: 139–159.

_____. 1988. "Überlegungen zur Sprach- und Wortkunstforschung jenseits des Strukturalismus." In: *Energeia und Ergon* (Festschrift E. Coseriu), Ed. H. Thun, II. Tübingen, Gunter Narr, 259–266.

_____. 1989. "Further reflections on the poetics of *Doktor Živago*: Structure, technique and Symbolism." In: *Boris Pasternak and His Times*, Ed. L. Fleishman. Berkeley: Berkeley Slavic Specialties, 284–314.

_____. 1990a. "A semiotic view of poetic language." *Znakolog* 2: 31–41.

_____. 1990b. "Semiotic modeling systems, primary and secondary," *Language Sciences* 12: 53–63.

_____. 1991a. "Ett livs historia—ett liv i historien: *Doktor Zjivago*." In: *Boris Pasternak och hans tid*, Eds. P. Alberg Jensen, P. A. Bodin, and N. Å. Nilsson. Stockholm: Almqvist & Wiksell International, 91–101.

_____. 1991b. Review of Grzybek 1989.

Burg, P. 1990. "Zeichen und Modell. Kritische Bemerkungen zur Diskretheit von Zeichen- und Modellkonzept bei Ju. M. Lotman und Ch. S. Peirce." *Znakolog* 2: 43–57.

Cassedy, S. 1991. "Pavel Florensky's philosophy of language: its contextuality and its context." *Slavic and East European Journal* 35: 537–552.

Edelman, G. M. 1992. *Bright Air, Brilliant Fire: On the Matter of the Mind*. New York: Basic Books.

Florenskii, P. A. 1988. "Antinomiia iazyka." *Voprosy iazykoznaniia* 6: 88–125.

Grzybek, P. 1989. *Studien zum Zeichenbegriff der sowjetischen Semiotik (Moskauer und Tartuer Schule)*. Bochum: Brockmeyer.

Ivanov, V. 1988. "O lingvisticheskikh issledovaniiakh P. A. Florenskogo." *Voprosy iazykoznaniia* 6: 69–87.

_____. 1991. "Self-portrait of a Russian semiotician in his younger and later years." In: *Recent Developments in Theory and History. The Semiotic Web 1990*, Eds. T. A. Sebeok and J. Umiker-Sebeok. Berlin–New York: Mouton, 3–43.

Jacob, F. 1982. *The Possible and the Actual*. Seattle: University of Washington Press.

Jakobson, R. 1960. "Closing statement: linguistics and poetics." In: *Style in Language*, Ed. T. A. Sebeok. Cambridge, MA/New York: Technology Press of MIT/John Wiley & Sons, 350–377.

_____. 1963. "Efforts toward a means-ends model of language in interwar continental linguistics." In: *Trends in Modern Linguistics*, Eds. C. Morhmann et al. Utrecht: Spectrum, 104–108.

_____. 1966. "Quest for the essence of language." *Diogenes* 51: 21–37.

_____. 1975. "Structuralisme et téléologie." *L'arc* 60: 3–8.

Jameson, F. 1974. *The Prison-House of Language: A Critical Account of Structuralism and Russian Formalism*. Princeton, N.J.: Princeton University Press.

Koschmieder, E. 1965. *Beiträge zur allgemeinen Syntax*. Heidelberg: Winter.

Krieger, M. 1992. *Ekphrasis: The Illusion of the Natural Sign*. Baltimore and London: The Johns Hopkins University Press.

Lerdahl, F., and R. Jackendorff. 1983. *A Generative Theory of Tonal Music*. Cambridge, MA and London: MIT Press.

Lotman, Yu. M. 1976. *Analysis of the Poetic Text*, Ed. and Tr. D. B. Johnson, with a bibliography . . . by L. Fleishman. Ann Arbor: Ardis.

______. 1990. *Universe of the Mind: A Semiotic Theory of Culture*, Intr. U. Eco, Tr. A. Shukman. London–New York: Tauris.

Shapiro, M. 1969. *Aspects of Russian Morphology: A Semiotic Investigation*. Cambridge, MA: Slavica.

______. 1983. *The Sense of Grammar: Language as Semeiotic*. Bloomington: Indiana University Press.

______. 1991. *The Sense of Change: Language as History*. Bloomington and Indianapolis: Indiana University Press.

On the European (Re)discovery of Shamans

CARLO GINZBURG

University of Bologna

I

IN A BOOK which appeared in Venice in 1565, later reprinted and translated many times—*La Historia del mondo nuovo* (*History of the New World*)—the Milanese Girolamo Benzoni described what he had seen in the course of his fourteen years of travelling in the "newly discovered islands and seas" beyond the Ocean. About the island of Hispaniola, he related the following:

"In this island, as also in other provinces of these new countries, there are some bushes, not very large, like reeds, that produce a leaf in shape like that of the walnut, though rather larger, which (where it is used) is held in great esteem by the natives, and very much prized by the slaves whom the Spaniards have brought from Ethiopia.

"When these leaves are in season, they pick them, tie them up in bundles, and suspend them near their fire-place till they are very dry; and when they wish to use them, they take a leaf of their grain (maize) and putting one of the others into it, they roll them round tight together; then they set fire to one end, and putting the other end into the mouth, they draw their breath up through it, wherefore the smoke goes into the mouth, the throat, the head, and they retain it as long as they can, for they find a pleasure in it, and so much do they fill themselves with this cruel smoke, that they lose their reason. And there are some who take so much of it, that they fall down as if they were dead, and remain the greater part of the day or night stupefied. . . . See what a pestiferous and wicked poison from the devil this must be. It has happened to me," Benzoni continues, "several times that, going through the provinces of *Guatemala* and *Nicaragua*, I have entered the house of an Indian

Elementa, 1993
Vol. 1, pp. 23–39
Reprints available directly from the publisher
Photocopying permitted by licence only

who had taken this herb, which in the Mexican language is called *tabacco. . . .*"

Following in the footsteps of the Russian Formalists, above all Shklovsky, we have learned to look for estrangement in the gaze of a savage, a child or perhaps even an animal: beings cut-off from the conventions of civil life, which they record with a bewildered or indifferent eye, thus indirectly pointing to their lack of meaning. Here we find ourselves confronted with a situation which is paradoxically reversed: the foreigner is the Milanese Girolamo Benzoni; those who complete the meaningless gesture of lighting a cigarette and smoking it are the savage Indians—a stand-in for ourselves, inhabitants of the civilized world. In the flight of Girolamo Benzoni ("and immediately perceiving the sharp fetid smell of this truly diabolical and stinking smoke, I was obliged to go away in haste, and seek some other place.") one is tempted to see the symbolic anticipation of a multi-secular, historical phenomenon: the withdrawal of the non-smokers before the advance—which has perhaps already reached its outer limit—of the army of tobacco smokers.

This Milanese traveller's page is one of innumerable testimonies of the encounter between Europeans and the disconcerting novelties of the Ocean beyond: animals, plants, customs. Today it's fashionable to analyze these documents through a very general category, that of an encounter with the Other: an expression with a somewhat metaphysical flavor which, however, suitably underscores the strict intersection within those relationships of natural and cultural otherness. A few pages after Girolamo Benzoni's invective against the effects of tobacco ("See what a pestiferous and wicked poison from the devil this must be") there follows a description of the way in which the plant was utilized by the indigenous doctors for therapeutic purposes. The invalid, "intoxicated" from the smoke, "on returning to his senses told a thousand stories, of his having been at the council of the gods and other high visions": the doctors then "turn the invalid round three or four times, rubbing his back and loins well with their hands, making many grimaces at him, and holding a pebble or bone in their mouth all the time. These things the women keep as holy, believing that they aid child birth. . . ."[1] It is clear that in the eyes of the Milanese traveller the indigenous doctors were simply witches: and the effects of the tobacco administered by them, mere diabolical hallucinations.

Although interlaced with considerations of opposite meaning, these negative connotations of tobacco are found in a book of a few years later, by the doctor from Seville, Monardes: *Primera y segunda y tercera partes de la Historia Medicinal.*[2] On the one hand, there are exaltations of the "great [curative] powers" of tobacco, recently introduced to the gardens and orchards of Spain, for every sort of illness: asthma, consumption, stomach and uterine pains. On the other hand, there are scandalized descriptions of the use which the Indians, in their religious ceremonies, made of that miraculous herb. Monardes writes that the priests, before divining the future, became dazed by the tobacco smoke until they fell onto the ground like dead men: then, once returned to their senses, they responded to the queries which were put to them, interpreting "in their own way, or following the inspiration of the Devil," the apparitions and illusions they had perceived in their cataleptic state. But the priests were not the only ones "to get drunk" (*emborracharse*) with the tobacco smoke: the Indians would do the same thing in order to bring pleasure or portents of the future from the images which presented themselves to their minds. "The Devil, who is a deceiver and knows the power of the herbs," Monardes comments, "has taught the Indians the virtue of tobacco: and leads them into deceit through visions and apparitions which the tobacco procures."

For Monardes, then, one of the characteristics of tobacco is the ability to procure "visions and apparitions" that the ancient doctors had attributed to the deadly nightshade, anise or horseradish.[3] But in Monardes' work, the more detailed comparison is provided by two substances, both endowed with hallucinatory powers and largely consumed in the Eastern Indies: *bangue*[4] and *anphion*—identifiable with marijuana and opium, respectively. As for *bangue* (or *Cannabis indica*, as it was classified by European botanists) Monardes cites and relies upon the pages pertaining to this plant by the Portuguese doctor Garcia da Orta, the author of a work, in the form of a dialogue, on the herbs and aromas of the Eastern Indies[5]: but he adds details and specifications based upon direct observation. Garcia da Orta spoke generically about the diffusion of *bangue* and opium; Monardes claims that the latter is preferred by the poor whereas the rich prefer *bangue* which is tastier and more aromatic. A few years before, the doctor of Burgos, Christoval Acosta, in his *Tractado de las Drogas, y medicinas de las Indias Orientales* had traced a topology of the consumers of *bangue*: some take it in order to forget fatigue and to sleep peacefully; others in order to amuse themselves in sleep with dreams and multi-colored illusions; others in order to get drunk; others for its aphrodisiac effects (something Monardes completely omits however); the great lords and captains to forget that which worries them.[6] All of the testimonies concur in underscoring how the inhabitants of the Eastern Indies had become accustomed to these narcotic substances: amazed, Monardes describes how five seeds of opium kill one of us; sixty seeds give them health and rest.

Rest, *descanso*: the barbarians of the Western Indies (it is still Monardes who is writing) resort to tobacco in order to overcome exhaustion; whereas those of the Eastern Indies resort to opium—in those parts a very common substance, which is sold in the shops.[7] In Peru, Girolamo Benzoni reports for his part, the natives "hold an herb in their mouth called *coca*, which must yield some nourishment, for they can walk a whole day without eating or drinking; this herb is their principal merchandize."[8] The meaning of these accounts, if examined in a broad, multisecular prospective, is very clear. The transoceanic voyages of exploration primed a circulation of intoxicating and narcotic substances so vast and intense that it could be compared to the microbic unification of the globe illustrated by Emmanuel Le Roy Ladurie in a famous essay.[9] In fact, in the span of a few centuries, tobacco, opium, marijuana and derivatives of coca penetrated (in various ways and to very different degrees) the culture of the colonizing peoples; wine and liquor, at a much faster pace, into the cultures of the colonized peoples.

I have touched on a tragic problem which is often addressed with irresponsible lightness. So I am anxious to immediately dispel a possible misunderstanding. Some have maintained that, since all of the intoxicating and narcotic substances are, insofar as they are drugs, potentially harmful, it is inevitable—if one does not wish to fall into an untenable, generalized prohibition—that their sale be legalized without exception. I would suggest that this conclusion is contrived; and the premise which must support it, false. Many, perhaps the majority of human societies, have used and continue to use, in very dissimilar forms and circumstances, substances which allow their users temporary access to a sphere of experiences which are distinct from the ordinary. Temporary escape (be it partial or total) from history is an indispensable ingredient of human history. But the degree of social control that each culture—besides of course, the single individuals within it—exercises with regard to these substances is extremely varied, and only partly explicable by a pharmaceutical analysis of their effects. Moreover, in each instance a filter, a cultural component whose function for the most part escapes us, interferes. How is it that alcoholic beverages, which for better or worse, European societies learned to live with over the course of several millennia (as in the case of wine) or several centuries (as in the case of liquors) had instead such a profoundly destructive effect upon the indigenous cultures of North America in the span of a few decades?

This is a well-known example. I mention it here because it allows me to introduce an extraordinary page taken from the account that the French Jesuit Paul de Brebeuf sent in 1636 to the provincial priest of the Company to report on the events which transpired in the Québec mission during that year. One of the priests had explained to the natives (naturally the account refers to them as "sauvages") that the high mortality rate they were suffering was due to wine and liquors which they didn't know how to use in moderation. "Why don't you write to your great King," one of the natives said, "that he prohibit these beverages from being brought in, since they kill us?" The Jesuit replied that the French needed these drinks in order to face the oceanic voyages and incredible cold of these regions. "Then make it such," the other replied, "that only they drink it." At this point, another native stood up: "No, it's not these beverages which take life away from us, it is your writings: since you first described our country, our rivers, our lands, our forests, all of us began to die, as did not occur before your arrival."

Paul de Brebeuf and his brethren responded to these words with loud laughter.[10] Today, with three and a half centuries behind us, we can admire the lucidity of the unknown native and at last grant that he was right. The Jesuits' cartography led the way in the European colonial conquest[11]; the immoderate assumption about the alcoholic beverages brought by the Europeans was nothing more than an aspect of the cultural decay of the natives produced by colonization.

Also the use of intoxicating, narcotic substances by the colonizers was similarly conditioned by filters which are cultural in character. But the way in which these filters acted is anything but obvious. Given the reactions of travellers, missionaries and botanists-doctors, had an imaginary sixteenth-century person tried to predict which of the intoxicating, narcotic substances would have arrived first for consumption in the Old Continent, presumably he would have bet on *bangue*, opium or coca. In fact, at least in the testimonies which I have examined, these substances are described in a neutral and detached tone, devoid of any moral or religious rebuke. On the other hand, tobacco—even in the pages of one who insists on its extraordinary medicinal powers like the doctor from Seville, Monardes—insistently calls to mind vice, sin or even the Devil. But these condemnations notwithstanding—or perhaps because of them—it was precisely the "pestiferous and wicked" tobacco which prevailed in Europe.[12]

II

Why did these intoxicating substances provoke such different reactions among sixteenth-century European travellers? A response to this question cannot be anything but cautious and provisional. As far as marijuana (or *bhang* if one prefers), opium and coca are concerned, we cannot rely on a historical, rational and systematic bibliography comparable to the five volumes *in folio* that Jerome E. Brooks started publishing in 1937 under the title, *Tobacco*.[13] Presumably the hypotheses which I am about to formulate are subject to correction on the basis of broader and more careful study.

I will begin with a few pages of a famous work: the *Historia general y natural de las Indias* by Gonzalo Fernandez de Oviedo (1535). The second chapter of the fifth book is concerned with tobacco use on the island of Española. From the beginning, Oviedo's voice resonates with a strong tone of moral reprobation: "The Indians of this island, besides their other vices, possessed one of the very worst (*muy malo*): that of inhaling smoke, which they call tabacco, for the purpose of losing their senses."[14] A description follows which concurs in many ways with that of Girolamo Benzoni (who, presumably, in preparing his own work, relied on Oviedo's account).[15] Oviedo observed that the Indians cultivated the plant, believing that its use was something "not only healthy but holy" (*no tan solamente les era cosa sana, pero muy sancta cosa*). They sometimes rely on this to alleviate their own physical ailments; some Christians follow this practice as well. The black slaves avail themselves of it to wipe out the exhaustion accumulated by the end of a day's work. But in the concluding paragraph, the condemnation again replaces these descriptive annotations:

"Here it seems appropriate to recall a vicious and evil custom which the people of Thrace practiced among their other criminal vices, according to what Abulensis writes about Eusebius *De observatione temporum* [III, 168] where he says that everyone, men and women, follow the custom of eating around a fire, attempting to be or appear drunk; and since they have no wine, they drink the seeds of certain herbs which grow in these parts and they throw them among the embers. They give off a smell which intoxicates all present even in the absence of wine. In my opinion, this is the same as the tabacco which the Indians use."[16]

Abulensis is the Spanish theologian Alonso de Madrigal, better known as Alonso Tostado, bishop of Avila. In his commentary on the *Ecclesiastical History*, printed in Salamanca in 1506, he speaks about the Thracian custom of gathering around a fire in order to get intoxicated with the smoke of certain baked seeds on the basis of the account provided by Solinus, the Latin cosmographer.[17] But Tostado's comment on the absence of wine among the Thracians comes from Solinus' source, the geographer Pomponius Mela. In the first century A.D., he prepared a work, *De orbis situ*, which, in the chapter dealing with Thrace, describes the ceremony with which we are already familiar.[18]

The story does not end here, because Pomponius Mela, in turn, had transferred onto the Thracians the description of a Scythian custom furnished by

Herodotus (IV, 73–75). But more about this later. First I would like to clarify the meaning of this digression. It has allowed us to reconstruct the cultural filter which permitted Oviedo (and not only him, as we shall see) to tame the natural and cultural otherness of the North American Continent. Thanks to Pomponius Mela and Solinus, the intoxicating herb smoked by the Indians came to be identified with that used by the Thracians, of uncertain nature but with equally intoxicating effects. This assimilation was facilitated by an obvious element: the inebriation brought on by alcoholic beverages and above all by wine, which constituted, as much as for a Latin cosmographer of the first century A.D. as for a French or Italian traveller fourteen or fifteen centuries later, the implicit model through which one could describe and evaluate the activity provoked by any intoxicating substance. Pomponius Mela observed that the Thracians, although unfamiliar with wine, entered a state of gaiety similar to drunkenness inhaling the smoke which is emitted from the baked seeds of an unspecified herb. In his history of Canada, the Jesuit François du Creux wrote in 1664 that the inhabitants of those lands always travelled fortified with *petun*, that is tobacco, and with a "fairly long tube" (a kind of pipe) so that they were able to enter a state of inebriation "analogous to that procured by wine."[19] It could very well be that this learned Jesuit, who was ready to compare the nomadism of the native Canadians with that of the Scythians,[20] was familiar with Pomponius Mela's passage. But the perception of tobacco as an alternative to wine goes far beyond an occasional erudite echo. This perception permeates the vocabulary of travellers and missionaries. Girolamo Benzoni wrote in passing about the natives of the Island of Hispaniola that "There are some who are happy to drink of this smoke." We read in an account prepared half a century later by the Jesuit Pierre Biard that the Canadian savages "also used petun [that is, tobacco] and they drank the smoke."[21]

That tobacco was used by the native North Americans on occasions ritual in character did not escape many European observers. It is again Pierre Biard who underscores the fact that among the savages of New France, any ceremony—from decisions to treatises, to public welcomes—called for the use of *petun*, or tobacco: "They place themselves in a circle around a fire, passing the pipe from hand to hand, and in this way they spend many pleasurable hours together."[22] The recognition of a ritual, if not utterly religious, dimension of the use of tobacco is also noted in Oviedo's words recalled earlier: for the natives of Hispaniola it is something "not only healthy but holy." The work which the priests performed on the same island has already been discussed. All of which suggests that tobacco, insofar as it was an instrument of private pleasure and public ritual, in the eyes of the European observers seemed like a wine which had lost its positive charge: a sort of sacred drink but employed by the natives in ceremonies which were considered idolatrous. From here the difference between the detached behavior adopted with respect to opium, *bhang*, and coca—intoxicating substances which the European observers associated, correctly or mistakenly, with a form of purely private consumption—and the manifest hostility for tobacco.[23] A hostility destined, however, to yield before the threatening offensive by producers of tobacco, cigars and pipes.[24]

At the beginning of the sixteenth century, one will recall, Oviedo had interpreted the Indian tobacco smokers through the texts about the Thracians by Pomponius Mela and Solinus. In the middle of the seventeenth century the situation reversed itself and the great erudite Isaac Vossius read an allusion to tobacco in Pomponius Mela's passage. Ivy, *arbutus* (strawberry tree) and cyclamen can produce intoxication: but what other herb, except tobacco ("praeter Nicotianam"), is capable of placing someone in a stupor with its own smoke?[25]

This rhetorical question took for granted that tobacco was already known in antiquity: a theory often advanced since the sixteenth century.[26] In 1724, the learned Jesuit Joseph-François Lafitau returned, with amplifications, to this discussion: an inevitable decision, in a work like his, entitled *Moeurs des sauvages amériquains, comparés aux moeurs des premiers temps* (1724).[27] As far as the Greeks and Romans were concerned, Lafitau came, perhaps somewhat reluctantly, to a negative conclusion. But Maxim of Tyre's passage on the Scythians,[28] as had those by Pomponius Mela and Solinus, seemed to him to constitute solid proof (even if not definitive) of the use of tobacco by those barbarian populations. It was an element which was added to innumerable others which, according to Lafitau, attested to the European origin of the first inhabitants of the American continent.[29] But the demonstration of the antiquity of the use of tobacco turned into a heated exaltation of its powers, significant because it had been prepared in resolutely non-Eurocentric terms, which meant that the earlier negative connotations were overturned. What had been in Europe a simple, unnecessary consumption, in America was (Lafitau emphasized) a holy herb "with multiple religious uses" (*à plusieurs usages de religion*). Besides the power which was attributed to it of "weakening the fire of concupiscence and the rebellion of the flesh, [tobacco serves] to illuminate the spirit, to purify it, to predispose it to ecstatic dreams and visions; it serves to evoke the spirits forcing them to communicate with men and to come in contact with the needs of the people who give them homage, it serves to cure all infirmities of the soul and of the body. . . ."[30]

III

Ecstatic dreams and visions, communication with spirits: as one can see, I am approaching the theme which I had announced in the title *On the European (Re)discovery of Shamans*. What I have said until now encompasses a series of merely apparent digressions, imposed by the fleeting nature of my topic.

During the years in which the Jesuit Lafitau printed the fruit of his grandiose and bold meditations on the customs of the American continent, the Russian expansion into Central Asia and the Far East was in full progress. Descriptions of those remote lands and nomadic populations which inhabited them, provided by the most varied characters, began to arrive in the West.[31] In 1704, the Dutch merchant E. Isbrants Ides, the ambassador of the Muscovite Czar, recorded for the first time the Tungusian term "schaman," as a synonym for sorcerer.[32] Shortly thereafter, the captain of dragoons, Johann Bernhard Müller, already in the service of the King of Sweden, and later a Russian prisoner of war, included in a description of the Ostiaks and their customs, an analytic account of a shamanistic seance (even if, presumably, it was not based on direct evidence), accompanied by catalepsies and divinations.[33] Towards the middle of the eighteenth century, the often massive works of scholars who had participated in their truly scientific expeditions in Siberia began to appear, like the one which lasted a total of ten years, composed of Johann Georg Gmelin, a professor of chemistry and botany in Tübingen; of the doctor Messerschmidt; of the philologist Müller and of the botanist Amman. In a prolix travel account in three volumes Gmelin described his encounters with the Tungus and Buriat shamans, who, in some instances, revealed their tricks to him.[34] It is clear that Gmelin considered these individuals coarse swindlers; nevertheless, he scrupulously transcribed their chants.[35] In addition, their ecstasies piqued his curiosity; in his great Latin work on Siberian flora he noted that the Buriats made use of juniper as a way to awaken their "praestigiatores" (that is, shamans) from the spell, and that the inhabitants of Kamchatka peninsula used nettle in their idolatrous cults, which was believed to be most suitable for procuring visions.[36]

Over the course of a few decades this convergence of research assured the Siberian shamans a prominent place in the panorama traced by the comparative science of religions, at that time in the process of defining itself.[37] In this sense, a meaningful example is provided by the short work by Meiners, a professor at Göttingen: *Grundriß der Geschichte aller Religionen* (Lemgo 1785). The title is deceptive: this is an early example of phenomenology, not of history of religions. The choice of an exposition "based on the natural succession of the most important constitutive elements (*nach der natürlichen Folge ihrer wichtigsten Bestandtheile*)" instead of being based on the chronological-geographical succession, took all religions into consideration *en masse*, revealed and otherwise, with obvious deistic implications. In the chapter significantly focusing on *Jongleurs* (literally, "jugglers") and priests, the shamans were assigned a niche apart.[38] But then the subdivision according to themes made the shamans reappear in the most unthinkable places: for example, at the end of the bibliographical notes to the chapter on sacrifice (including also human sacrifices), which begin with the Pentateuch, continue with Greek and Roman authors, and end with a contemporary traveller, J. G. Georgi, the author of a descriptive account of Siberia.[39]

Ten years earlier, Meiners had published a vast essay, "On the mysteries of the ancients, and in particular on the Eleusinian secrets," which was preceded by an introduction written in a comparative perspective.[40] In the latter he distinguished between the mysteries performed by priests and mysteries attached to oral or written doctrines: in neither of the two cases could the phenomenon be considered universal. It was unknown to the Samoyeds, to the inhabitants of the Kamchatka peninsula, to the Tartarian hordes (about which Meiners deferred to Gmelin), to the Californians, to the Eskimos, to the Laplanders and to the Greenlanders. Meiners observed that in all of these populations, one could not speak of common religion or of national gods, nor even of priests in a strict sense, but only of "charlatans and soothsayers" (*Quacksalber und Wahrsager*).[41] In this way shamans, even if they were merely evoked and not named directly, steadily entered the religious history of humanity in order to mark its poorest, most elementary stage.

IV

Thus, thanks to the Russian Empire's expansion towards the East, Europeans discovered shamans. They discovered, or better yet, they rediscovered. This clarification seems to me appropriate for two reasons. First of all, between the sixteenth and seventeenth centuries erudites like Peucer and Scheffer had collected and conveyed information about Laplander enchanters, closely related to (as Meiners well understood) the Siberian shamans.[42] Secondly, as I have tried to demonstrate more fully elsewhere, an extremely ancient shamanistic nucleus was contained in the well-known stereotype of the witch's sabbath.[43]

Cognition—recognition—are complicated operations. Perceptions and cultural schemata intersect, conditioning one other in turn. For Monardes, the doctor from Seville, the Indian priests who began to foretell the future after leaving the cataleptic state brought on by the tobacco smoke were inspired by the Devil. A great erudite like Vossius recognized in Pomponius Mela's Thracians tobacco smokers. Clearly, Vossius was mistaken (and with him, Lafitau). He was, however, absolutely correct in recalling the description of a Scythian rite taken from the fourth book of Herodotus.[44] Here it is:

"After the burial the Scythians cleanse themselves as I will show: they anoint and wash their heads; as for their bodies, they set up three poles leaning together to a point and cover these over with woollen mats; then, in the place so enclosed to the best of their power, they make a pit in the centre beneath the poles and the mats and throw red-hot stones into it.

"They have hemp growing in their country, very like flax, save that the hemp is much the thicker and taller. This grows both of itself and also by their sowing, and of it the Thracians even make garments which are very like linen. . . .

"The Scythians then take the seed of this hemp and, creeping under the mats, they throw it on the red-hot stones; and, being so thrown, it smoulders and sends forth so much steam that no Greek vapour-bath could surpass it. The Scythians howl in their joy at the vapour-bath. This serves them instead of bathing, for they never wash their bodies in water."[45]

The more or less analogous passages of Maxim of Tyre, Pomponius Mela and Solinus, which refer, respectively, to the Scythians (the first) and the Thracians (the last two), are taken from this page of Herodotus. This constitutes an historical document of enormous importance. As far as I know, the first step towards the correct interpretation of it was made by an antiquarian-naturalist, Engelbert Kaempfer (1651–1716).[46] The illustrations which accompany the collection of observations accumulated over the course of years and years of travel—*Amoenitatum Exoticarum politico-physico-medicarum fasciculi V*—give an idea of Kaempfer's boundless curiousity: one moves from an inscription in cuneiform characters transcribed from the ruins of Persepolis to an exceedingly accurate description of the pressure points Japanese acupuncturists used to cure colic diarrhea.[47] One observation discusses the properties and effects of tobacco, opium and *cannabis*, or *bangue*: in the latter, Kaempfer identified the intoxicating plant with the Scythian and Thracian smoke.[48]

These lines apparently passed unnoticed. At the end of the eighteenth century another and no less extraordinary character, Count Jan Potocki—the author of *Manuscript Found at Saragossa*, the novel that an abridged version edited by Roger Caillois made famous the world over—independently arrived at similar conclusions.[49] In a splendid book which appeared in St. Petersburg in 1802, the *Histoire primitive des peuples de la Russie*, Potocki closely examined the customs of the nomadic populations of Central Asia described in the fourth book of Herodotus. He unhesitatingly recognized "les Schamanes de la Sibérie" in the Scythian soothsayers.[50] Among the Tartar populations, he had not located the custom of entering a trance from the smoke of baked hemp seeds: but he noticed that *haschisch*, widely used in Cairo (where he had spent some time in 1790) yielded a state of drunkenness which differed from that of opium and fermented liquors, because "tient davantage de la folie."[51]

These intuitions also passed unnoticed. In an essay read in 1811 and later revised for publication in 1828, Niebuhr, the great German historian, skillfully traced the most ancient historical lineages of the Scythians, Getae and Sarmatians, arriving at closely similar conclusions to those of Potocki but without mentioning him, because he clearly had no knowledge of his work. In the funeral ceremony described by Herodotus (IV, 73–75) Niebuhr unreservedly saw a shamanistic ritual, that supported the hypothesis which he maintained, and is still discussed today, for a Mongolian origin of part of the Scythian populations.[52]

The convergence between Potocki and Niebuhr on this specific point is not casual. In a course on Slavic literature held at the Collège de France in 1842–1843, Adam Mickiewicz said that Potocki had been "the first among historians of modern Europe to recognize the importance of the oral tradition. Niebuhr asked the peasants and old women in the Roman markets for explanations of the story of Romulus and Remus. Many years before him, Potocki had meditated on the history of Scythians in the Tartar tents." And he concluded: Potocki had travelled, observed the places, spoken with the people–things no antiquarian had done before him.[53]

Mickiewicz, understandably, exaggerated: one has only to think of the travels undertaken, towards the end of the seventeenth century, by the antiquarian and naturalist Engelbert Kaempfer. But he was indeed correct in emphasizing the richness of an approach which was rediscovered by ethnographers in recent decades. The classicist and folklorist Karl Meuli has explored this avenue, and in an essay which appeared in 1935 (*Scythica*) rediscovered, perhaps for the last time, the shamanistic connotations of the Scythian funeral rite described by Herodotus. I say "rediscovered" because, if I am not mistaken, in the extensive notes which accompany Meuli's article, the names of scholars who had anticipated his basic thesis to a greater or lesser degree, are missing: Kaempfer, Potocki and (curiously enough)

Niebuhr. In any event, this does not take away from the originality of Meuli's splendid essay, which for the first time analyzed in depth both the shamanistic elements present in the Scythian culture as well as their absorption by the Greek colonists settled along the banks of the Black Sea.[54]

Unbeknownst to Meuli, the results of an archaeological excavation completed a few years earlier in the eastern Altai Mountains had provided unexpected evidence which anticipated the essay's conclusions. In a place called Pazyryk some tombs had been found, dating from the second or third century B.C., in which the following remains were found: a horse disguised as a reindeer (now on display at the Hermitage); a drum similar to those used by shamans; a few seeds of *cannabis sativa*, some preserved in a leather pouch and some baked among the stones contained in a small, bronze washbasin.[55]

V

I believe that the accumulation of knowledge always happens in this way: across broken rather than continuous lines; through false beginnings, corrections, oversights, and rediscoveries; thanks to filters and schemata which blind and, at the same time, illuminate. In this sense the case which I have reconstructed in perhaps excessive detail can be considered almost banal: not the exception but the rule.

Notes

1. See G. Benzoni, *La historia del mondo nuovo*, Venezia 1565, pp. 54v–55v (=*History of the New World* tr. W. H. Smyth, New York 1857 [19—], pp. 80–81); the citation is verified in the second edition, partially republished by A. Martinengo, in *Nuovo mondo. Gli italiani (1492–1565)*, edited by P. Collo and P. L. Crovetto, Torino 1991, p. 576 (see also Martinengo's note to pp. 549–552). See also the entry prepared by A. Codazzi for the *Dizionario biografico degli italiani*, vol. 8, pp. 732–733, who mentions part of the passage reproduced above. For a picture of the reprintings and traditions, see the anastatic reprint of the second edition of Benzoni's work edited by F. Anders, Graz 1962.
2. See N. Monardes, *Primera y segunda y tercera partes de la historia Medicinal, de las cosas que se traen de nuestras Indias occidentales, que sirven en Medicina*, Sevilla 1580 (1st ed. 1571), cc. 32r sgg., especially 36v–39r. See also N. A. Recco, *Rerum medicarum Novae Hispaniae Thesaurus. . .*, Romae 1648, pp. 173–177 (I.V, ch.LI: "De Pycielt, seu Tabaco").
3. See P. A. Mattioli, *I discorsi . . . nelli sei libri di Pedacio Dioscoride Anazarbeo della materia medicinale*, Venezia 1568, p. 1476, on the "Solatro maniaco over furioso," which is distinguished from "doricnio" (p. 1132: Mattioli says that he was unable to identify it) which is also mentioned by Dioscorides.
4. According to contemporary transcriptions, *bhang*.
5. Garcia da Orta, *Coloquios dos simples e drogas da India*, Lisboa 1891, annotated by conte de Facalho, pp. 95–101. The first edition, which I have not seen, appeared in Goa in 1563.

6. See C. Acosta, *Tractado dellas drogas, y Medicinas de las Indias Orientales, con sus plantas debuxadas al bivo . . .* Burgos 1578, pp. 360–361. In the dedication to the reader Acosta discretely points out the faults of his predecessor's (Garcia da Orta) work.
7. See N. Monardes, *Primera* cit., c. 38 r.
8. See G. Benzoni, *La Historia* cit., p. 167v (Eng. tr. p. 249).
9. See E. LeRoy Ladurie, "Un concept: l'unification microbienne du monde (XIVe–XVIIe siécles)," in *Le territoire de l'historien*, II, Paris 1978, pp. 37–97.
10. [Père Paul de Brebeuf], *Relation de ce qui s'est passé en la Nouvelle France en l'annéee 1636, envoyée au R. Père Provincial de la Compagnie de Jésus en la Province de France par le P. Paul le jeune de la mesme Compagnie, supérieur de la Résidence de Kébec*, Paris 1637, I, pp. 199–200: ". . . monsieur Gand parlant aux Sauvages, comme j'ay dit cy-dessus, leur remonstroit, que s'ils mouroient si souvent, ils s'en falloit prendre à ces boissons, dont ils ne scauroient user par mesure. Que n'écris tu à ton grand Roi, firent-ils, qu'il défende d'apporter de ces boissons qui nous tuent. Et sur ce qu'on leur repartit, que nos François en avoient besoin sur la mer, et dans les grandes froidures de leur païs, fais donc en sorte qu'ils les boivent tous seuls. On s'efforcera, comme j'espère, d'y tenir la main; mais ces Barbares sont importuns au dernier point. Un autre prenant la parole, prit la defense du vin et de l'eau de vie. Non, dit-il, ce ne sont pas ces boissons qui nous ostent la vie, mais vos écritures: car depuis que vous avez décry nostre païs, nos fleuves, nos terres, et nos bois nous mourons tous, ce qui n'arrivoit pas devant que vous vinssiez icy. Nous-nous mismes à rire entendans ces causes nouvelles de leur maladies. Je leur dy que nous decrivions tout le monde, que nous décrivions nostre païs, celuy des Hurons, des Hiroquios, bref toute la terre, et cependant qu'on ne mouroit point ailleurs, comme on fait en leurs païs, qu'il falloit donc que leur mort provint d'ailleurs; ils s'y accordèrent." Lafitau reports that Brebeuf's priest died, killed by the Iroquois (II, 131: cf. *Dictionnaire de biographie française, ad vocem*).
11. See F. de Dainville, *La géographie des humanistes*, Paris 1940 (new ed., 1991).
12. On the hostility (which lasted until the end of the seventeenth century) of the Catholic hierarchy with respect to tobacco consumption, see J. Tedeschi, "Literary Piracy in Seventeenth Century Florence: Giovanni Battista Neri's *De iudice S. Inquisitionis Opusculum*". In Id. *The Prosecution of Heresy. Collected Studies on the Inquisition in Early Modern Italy*, Binghamton 1991, pp. 259–272.
13. See J. E. Brooks, *Tobacco. Its History Illustrated by the Books, Manuscripts and Engravings in the Library of George Arents, Jr.* 5 vols, New York, 1937, sgg.
14. See G. F. de Oviedo, *Historia general y natural de las Indias*, edited by J. Perez de Tudela Bueso, I, Madrid 1959, pp. 116–118.
15. Benzoni's words "fumo che in lingua Mesicana è chiamato tabacco," and later corrected in the second edition to "quest'erba che ecc. . ," could have been suggested by Oviedo's observation: the Indians call the smoke or pipes for inhaling tabacco, not (as some believed) the herb or sleep in which they fell after having smoked it (*Historia* cit., p. 116). Instead, successive authors, like for example Monardes, call the plant "tabacco" according to a custom later invalidated. According to A. Ernst, "On the Etymology of the Word Tobacco" in *American Anthropologist*, II (1889), pp. 133–141, the instrument described and reproduced by Oviedo—in the Guaraní language, *taboca* had been and continued to be used on the American continent for inhaling not tobacco smoke but alkaloid-containing weeds. The hypothesis that Benzoni never actually made the voyages he described is discussed (and repudiated) by A. Codazzi and A. Martinengo cited in note 1.
16. "Aquí me paresce que cuadra una costumbre viciosa e mala que la gente de Tracia usaba entre otros criminosos vicios suyos, segundo el Abulensis escribe sobre Eusebio *De los tiempos* [lib. III, cap. 168], donde dice que tienen por custumbre todos, varones e mujeres, de comer alrededor del fuego, y que huelgan mucho de ser embriagos, o los parescer; e que como no tienen vino, toman simientes de alguanas hierbas que entre allos hay, las

cuales, echadas en las brasas, dan de sí un tal olor, que embriagan a todos presentes, sin algo beber. A mi parescer, esto es lo mismo que los tabacos que estos indios toman" (G. F. de Oviedo, *Historia* cit., I, p. 117).

17. See *Tostado sobre el Eusebio*, Salamanca, Hans Gysser 1506, t. III, c. lix v (cap. clxviii); C. Julius Solinus, *Polyhistor, rerum toto orbe memorabilium thesaurus locupletissimus*, Basileae 1538, p. 36: "Uterque sexus epulantes, focos ambiunt, herbarum quas habent semine ignibus superiecto, cuius nidore perculsi, pro laetitia habent, imitari ebrietatem sensibus sauciatis."
18. See Pomponio Mela, *De orbis situ libri tres, accuratissime emendati, una cum commentariis Joachimi Vadiani . . .*, Parisiis 1540, p. 90: "Vini usus quibusdam ignotum est: epulantibus tamen ubi super ignes, quos circunsident, quaedam semina ingesta sunt, similis ebietati hilaritas ex nidore contingit."
19. François du Creux, *Historiae Canadensis, seu Novae Franciae libri decem*, ad annum usque Christi MDCLVI, Parisiis 1664, p. 76 ". . . ebrietatemque enim inducunt, vini instar" (before the same page there is an illustration which shows a native who is smoking a pipe).
20. Ibid., p. 56.
21. See P. Biard, Grenoblois de la Compagnie de Jésus. *Relation de la nouvelle France, de ses terres, naturel du Païs et de ses Habitans . . .*, Lyon 1616, p. 78.
22. Ibid., pp. 78–79.
23. On the use of *bhang* in ritual contexts, see R. G. Wasson, *Soma, Divine Mushroom of Immortality*, New York, 1971, p. 128 sgg., which discusses, and rejects, the proposal to identify *bhang* with the *soma* mentioned in Vedic poems (regarding B. L. Mukherjee, "The Soma Plant," *Journal of the Royal Asiatic Society*, 1921, pp. 241–244; a booklet by the same author, with the same title appeared in Calcutta in 1922, I have not seen it).
24. See S. Schama, *The Embarrassment of Riches*, New York 1987, p. 193 sgg.
25. See I. Vossius, *Observationes and Pomponium Melam de situ orbis*, Hagae Comitis 1658, pp. 124–125.
26. The oldest representation of tobacco by a European botanist (the Dutchman Rembert Dodoens, 1554) identifies the plant with the *Hyoscyamus luteus* described by Dioscorides: See J. Stannard, "Dioscorides and Renaissance Materia Medica," in *Analecta medico-historica*, I (1966), *Materia medica in the Sixteenth Century*, edited by M. Florkin, p. 133 (and n. 93); F. Edelemann, "Nicotiniana," extracted from *Flammes et Fumées*, 79 (1977).
27. See J.-F. Lafitau, *Moeurs des sauvages amériquains, comparés aux moeurs des premiers temps*, II. Paris 1724, pp. 126 sgg., about whom see A. Pagden, *The Fall of Natural Man*, Cambridge 1982; (*La caduta dell'uomo naturale*, Italian trans., Torino 1989, pp. 256–270).
28. See Maxim of Tyre, *Sermones sive disputationes XLI*, Parisiis 1557, p. 90 (serm. XI).
29. See J.-F. Lafitau, *Mémoire présenté au Duc d'Orléans concernant la précieuse plante de Gin Seng de Tartarie découverte au Canada*. Paris 1718 (which I have not seen).
30. See J.-F. Lafitau, *Moeurs*, cit., II, p. 133: "Il est certain que le Tabac est en Amérique une herbe consacrée à plusieurs exercices, et à plusieurs usages de Religion. Outre ce que j'ai déjà dit de la vertu qu'ils lui attribuent pour amortir le feu de la concupiscence et les révoltes de la chair; pour éclairer l'âme, la purifier, et la rendre propre aux songes et aux visions extatiques; pour évoquer les esprits, et les forcer de communiquer avec les hommes; pour rendre ces esprits favorables aux besoins des nations qui les servent, et pour guérir toutes les infirmités de l'âme et du corps. . ."
31. See G. Henning, *Die Reiseberichte über Sibirien von Heberstein bis Ides*, "Mitteilungen des Vereins für Erdkunde zu Leipzig," 1905, pp. 241–394.
32. See E. I. Ides, *Voyage de Moscou à la Chine*, in *Recueil de voiages au Nord, contenant divers mémoires très utiles au commerce et à la navigation*, VIII, Amsterdam 1727 [in the Bibliothèque Nationale's catalogue in Paris, this work is catalogued under the name of its

editor, Jean-Frédéric Bernard] p. 54: "A quelques journées de chemin d*Ilinskoi* il y a une grande cascade, ou pente d'eau, qu'on appelle *Chute du Schaman*, ou *Chute du Magicien*, à cause que la fameux *Schaman*, ou magicien des *Tunguses*, a sa cabane auprès de cet entroit." The original Dutch version of Ides' account appeared in Amsterdam in 1704. On the term "shaman" see S. M. Shirokogoroff, *Psychological Complex of the Tungus*, London 1935, pp. 268–269 which also defers to B. Laufer, "Origin of the Word Shaman," in *American Anthropologist*, 19 (1917), pp. 361–371.

33. See Johann Bernhard Müller, *Les moeurs et usages des Ostiackes et la manière dont ils furent convertis en 1712 à la religion chretienne du rit grec*. In *Recueil de voiages du Nord*, VIII, Amsterdam 1727, p. 382 sgg., especially p. 412 (it is the translation of a German version which I have not seen).
34. See J. G. Gmelin, *Reise durch Sibirien, vor dem Jahr 1733 bis 1743*, 3 vols., Göttingen 1751–1752; see especially I, pp. 283 sgg., 351, 397; II, pp. 45–46, 82 sgg., 351; III, preface, pp. 69 sgg., 330 sgg., 347 sgg. A radically abbreviated French translation of this work exists: *Voyage en Sibérie*, 2 vols, Paris 1767.
35. See J. G. Gmelin, *Reise*, cit., III, pp. 370 sgg., 522 sgg.
36. See J. G. Gmelin, *Flora sibirica sive historia plantarum Sibiriae*, I, Petropoli 1747, p. 184; t. III, Petropoli 1748, p. 31. A biography of Gmelin, prepared by the Rector of the University of Tübingen, is prefaced to J. G. Gmelin, *Sermo academicus de novorum vegetabilium post creationem divinam exortu*, Tubingae 1749.
37. See F. E. Manuel, *The Eighteenth Century Confronts the Gods*, Cambridge, Mass. 1959.
38. Ibid., pp. 137–145.
39. Ibid., pp. 73–74. See J. G. Georgi, *Bemerkungen einer Reise im russischen Reich im Jahre 1772*, 2 vols., St. Petersburg 1775.
40. See C. Meiners, "Ueber die Mysterien de Alten, besonders über die Eleusinischen Geheimnisse," in *Vermischte Philosophisce Schriften*, III, Leipzig 1776, pp. 164–342. S. Landucci has recalled attention to Meiners (barely mentioned by Manuel) in *I filosofi e i selvaggi, 1580–1780*, Bari 1972, pp. 463–465 and *passim*; more generally, iinsisting on the Eurocentric and racist perspectives of his writings, L. Marino, *I maestri della Germania, Göttingen 1770–1820*, Torino 1975, pp. 103–112.
41. See C. Meiners, "Ueber di Mysterien," cit., pp. 169–171.
42. See C. Peucer, *Commentarius de praecipuis generibus divinationum*, Witebergae 1560; J. Scheffer, *Lapponia*, Francofurti and Lipsiae 1674.
43. See C. Ginzburg, *Storia notturna. Una decifrazione del sabba*, Torino 1989 (=*Ecstasies. Deciphering the Witches' Sabbath*, New York 1991).
44. See I. Vossius, *Observationes*, cit., p. 124.
45. Herodotus, IV, 73–75 (=*Histories*, A. D. Godley, II, Cambridge and London 1950, pp. 273–274).
46. See K. Meier-Lemgo, *Engelbert Kämpfer, der erste deutsche Forschungsreisende 1651–1716*, Hamburg 1960. See also, edited by the same Meier-Lemgo, "Die Briefe Engelbert Kaempfers," *Akademie der Wissenschaften und der Literatur, Abhandlungen der mathematisch-naturwissenschaftliche Klass*, Jahrgang 1965, nr. 6, pp. 267–314; *Die Reisetagebucher Engelbert Kaempfers*, Wiesbaden 1968.
47. See E. Kaempfer, *Amoenitatum Excoticarum politico-physico-medicarum fasciculi V*, Lemgoviae 1712, pp. 333–334, 528–529. See also L. Gwei-Djen and J. Needham, *Celestial Jewels*, Cambridge, 1980 (*Aghi celesti. Storia e fondamenti razionali dell'agopuntura e della moxibustione*, tr. it., Torino 1984, p. 338 sgg.). His manuscripts (among which there is an unedited account of his travels) are to be found in the British Library.
48. See *Amoenitatem*, cit., p. 638 sgg., in particular p. 647. Kaempfer's source is Alessandro d' Alessandro, *Genialium dierum libri sex*, Parisiis 1561, pp. 137v–138r (1. III, XI).
49. A new unabridged version exists edited by R. Radrizzani: *Manuscrit trouvé à Saragosse*, Paris 1990.

50. See J. Potocki, *Histoire primitive des peuples de la Russie,* St. Petersburg 1802, p. 128. The word "chaman" is used in the reprinted edition by the orientalist Julius Klaproth, a disciple of Potocki, and includes an introduction and important critical notes.
51. See J. Potocki, *Histoire,* cit., p. 134; Id., *Voyages en Turquie et an Egypte, en Hollande, au Maroc,* edited by F. Beauvois, Paris 1980 (with a helpful introduction). I wonder whether a secret relationship exists, especially in terms of a structural order, which has always seemed fleeting to me between Potocki's *Manuscrit* and Sadig Hideyat's *The Blind Owl* (New York 1990). It should not be looked for in a re-interpretation in a very different key from an analogous hallucinatory experience. On this quite remarkable writer, see Y. Ishaghpour, *Le tombeau de Sadègh Hedayàt,* Paris 1990.
52. See B. G. Niebuhr, "Untersuchungen über die Geschichte der Skythen, Geten, und Sarmaten (Nach einem 1811 vorgelesenen Aufsatz neu gearbeitet 1828)," In: *Kleine historische und philologische Schriften,* Bonn 1828, I, pp. 352–398, especially pp. 361–362.
53. See A. Mickiewicz, *L'Eglise officielle et le messianisme,* 2 vols., Paris 1845, I, *Cours de littérature slave au Collège de France (1842–1843),* pp. 123–125: ". . . le premier de tous les historiens de l'Europe moderne, il reconnut l'importance de la tradition orale. Niebuhr demandait aux paysans et aux vieilles femmes, sur les marchés de Rome, des explications sur l'histoire de Romulus et de Rémus. Longtemps avant lui, Potocki, dans les huttes de Tartares, méditait sur l'histoire des Scythes . . . Potocki le premier a tiré la science du cabinet. Il a voyagé, observé le pays, parlé avec les peuples, ce qu'aucun antiquare n'avait fait avant lui . . ." The passage is mentioned by E. Krakowski, *Un Témoin de l'Europe des Lumières: le comte Jean Potocki,* Paris 1963, p. 149. On the importance Niebuhr attributed to oral tradition, see A. Momigliano, *Perizonio, Niebuhr e il carattere della tradizione romana primitiva.* In: *Sui fondamenti della storia antica,* Torino 1984, pp. 271–293.
54. See K. Meuli, "Scythica," In: *Gesammelte Schriften,* edited by T. Gelzer, Basel-Stuttgart 1975, pp. 817–879 (with additions with respect to the version published in 1935). See also by this author, *Storia notturna,* cit., p. 198 note 4.
55. See *Storia notturna,* cit., p. 188 (=*Ecstasies,* cit., p. 208).

THE AVANT-GARDE AND SLAVIC TRADITIONS

Trans-Sense as a Signifier and a Signified in Non-Futurist Texts

OMRY RONEN

University of Michigan, Ann Arbor

ONE OF THE KEY CONCEPTS of avant-garde verbal art in Russia, *zaum'* is a terminological neologism coined by futurist poets and theoreticians and variously rendered in English as 'trans-sense,' 'transrational,' 'metalogical,' or 'metaintellectual poetic language.'[1] In his commentaries to G. O. Vinokur's essay on Khlebnikov, M. I. Shapir has recently suggested that the noun *zaum'*, in contradistinction to the adjective *zaumnyi*, was used, as a rule, by Khlebnikov's critics, while the poet himself preferred the term *zaumnyi iazyk* 'trans-sense language.'[2] Indeed, Khlebnikov's word for trans-sense as such, not as an attribute of language, is *zaum* rather than *zaum'*. It occurs, on a single occasion, in *Zangezi*. However, Kruchenykh, the most persistent partisan of trans-sense and the probable inventor of the term, did use the noun *zaum'*, beginning with his "Declaration of the Trans-Sense Language" (Baku 1921).[3] Obviously, the history of this usage within the avant-garde requires further study.

In any case, the term has become a shibboleth of Russian futurism and gained wide currency outside its limits. Its felicitous association with the idiomatic expression *um za razum zakhodit* 'the mind boggles' accounted for its general acceptance and even great popularity, albeit in a pejorative sense, among the philistines, whereas its recondite anagrammatic affinity with the word *muza* 'muse' was appreciated by the cognoscenti and became part of the neoteric secret lore.

Both connotations were eventually made advantage of by Nabokov, when he wrote of "special poetic meaning (when one's mind, after going around itself in the subliminal labyrinth, returns with newfound music that alone makes poems what they should be)": "tot osobyi poeticheskii smysl

Elementa, 1993
Vol. 1, pp. 43–55
Reprints available directly from the publisher
Photocopying permitted by licence only

(kogda za razum zashedshii um vozvrashchaetsia s muzykoi), kotoryi odin vyvodit stikhi v liudi."[4]

Although the actual spread of the phenomenon covered by the term in its strict sense[5] remained narrow, it was recognized as the purpose of futurist sound arrangement and verbal patterning, and theoretically identified, by Roman Jakobson, with the ultimate limit toward which poetic language tends in general: the purely euphonic word with suppressed referentiality.[6]

For those who did not belong to the self-avowed avant-garde, the concept represented a challenge which produced multifarious response, sympathetic or polemical, not only in theoretical **and** critical writing, but in poetry and fiction, ranging from metapoetic images of trans-sense to its occasional actual use.

In historical retrospect, two correlated sets of synthetic tropes combining mythological metonymies and metapoetic metaphors, with an underlying common meaning of 'shagginess,' appear prominently among the figurative representations of trans-sense. One set, the images of fur, animal hide, wool, tousled hair, and so on, is common to a number of poets. The other, which comprises references to hay, straw, herbs, moss, and so on, in sustained specific relation to trans-sense, is characteristic individually of the poetry and essays by Mandel'shtam, with some intertextually relevant cognates in Esenin's "Kobyl'i korabli' ["The Mares' Ships," 1919] and Tsvetaeva's "Molv'" ["The Talk," 1924].

The group of the images of feral fur, goat feet, fleece, and the like pertains to the anti-Apollinian, heliomachic theme of Russian futurism, formulated in Kruchenykh's opera *Pobeda nad solntsem* [*Victory over the Sun*, 1913] as an onslaught upon the aesthetic norm associated with 'the sun of Russian poetry,' Pushkin. The source of these images can be traced back to the story of Marsyas in the modern poetic interpretations largely inspired by Nietzsche's idea of the Dionysiac art born of the spirit of music. In 1911, Benedikt Livshits published a collection of poems entitled *Fleita Marsiia* [*The Flute of Marsyas*]: a year later, Henri de Régnier's long poem "The Blood of Marsyas" appeared in the Russian translation of Maksimilian Voloshin. The founder of acmeism Nikolai Gumilev contributed his share to the allegory of the creative regression of poetry freed from any aspirations beyond itself by reviving in his short verse drama *Akteon* [*Actaeon*, 1913] the myth of Diana's revenge (it will be recalled that the Italian futurist antecedent of the Russian heliomachic theme was Marinetti's "Let Us Murder the Moonshine," 1909, to which the Russian avant-garde responded with the miscellany entitled *Dokhlaia luna* [*The Carrion Moon*, 1913][7]). In Gumilev's play, the hunter and poet Actaeon refuses to carry stones for the temple his father builds:

I razve nadoben bogam,
V ikh radostiakh raznoobraznykh,
Toboiu vystroennyi khram [. . .]
Oni ne liubiat nashikh stran,
I nichego im zdes' ne nado.
A esli nado, to rodnik,
Lepechushchii temno i stranno [. . .]

Why should the gods have any use,
in their manifold joys,
for the temple you have erected? [. . .]
They love not our lands
and need nothing here.
Or, if they do, it is a water source
that babbles obscurely and strangely [. . .]

Father Cadmus responds:

Tak bud' schastliv,
Kak travy, mozhet byt', kak zveri.

Be happy then
as the grasses are and, may be, the beasts.

In fulfillment of these words the visionary who has peeked at the naked moon goddess is turned into a stag, and his body is discovered "wooly and horned, / worse than the lowliest of satyrs" ("v shersti i s rogami, / khuzhe samogo poslednego satira").

In 1917, when Khlebnikov had already greeted the new sunrise and proclaimed "general repairs of the sun's loose constellation,"[8] Andrei Belyi, in whose poetry and "symphonic" prose the goat dance (*kozlovak*) of grotesque theriomorphic characters had commenced long before the rebirth of Marsyas in futurist art, published a remarkable essay, "Aaron's Rod,"[9] in which he brought forward a new metapoetic myth expressing his growing dismay at the disruption of the sound-sense nexus in the poetic word. For the lawful word [*zakonnoe slovo*], which manifests to us the spiritual nature of consciousness, Belyi wrote, all the sounds of the purely acoustic natural word [*prirodnoe slovo*] are fauns: "Here the futurism of language induces panic fear." "The logical Logos fights here the forest Pan, the nature of the sounding word."

Belyi outlined the two possible paths of reconciliation between "futurism" and "logism" "in our days." One path is to deepen the antinomy of the words to the point of recognition that the place of logic pertains to the more fundamental plane than the one its immanent being occupies (logic, Belyi says elsewhere in this essay, is not the unity of reason but 'the particular Name' [*Imia Rek*], i.e., the name of Logos in the anthroposophic sense) and that the phonic value [*zvukovaia znachimost'*] is not the form of the 'vocal sound' but its meaning. This meaning, however, is extra-rational [*vnerazumen*] in poetry: "it is present in the dark chaos of the words which constitute for us the form. This dark nature of the word, the elemental force of the word [*stikhiia slova*] is the loud sound which rises against the naked abstraction of thought; a goat-footed Pan, it attacks the philosopher. Apollo's world of the word is broken [*Apollonov mir slova—sloman*]; onomatopoeia broke away from thought. . . ." This is, according to Belyi, the reason why some former adepts of the poetic speech and poetic thought have been captivated by the aesthetics of the terminology and methodology of logic whereas "others are now attracted not by the ability to read the message of the Parcae's babble but by the very process of babbling."[10]

Hence the other path advocated by Belyi is the path of unity, regained in the expected inspiration (in the anthroposophic sense) of the "inner-born Word." In search of "poetry's inspirational root" in Logos, Belyi examines analytically the sound structure of poetry to find in it the higher "logic of alogism," "inaccessible to our logic" but "sensed in the cosmic rustle of phonetical effluences." The new logic is the logic of the identity of sound and sense, which Belyi discovers in the orchestration of poetry [*instrumentovka*]: "The orchestration of poetry is a sunnily-rational speech; it has its own signs, and these signs have not yet been read. . . ." "The enterprising flexibility in the quest for correspondence between sound and thought," understood as "the ability to read the *new word* in a worn-out word," is for Belyi the pledge of the birth of "baby Hermes" to replace the goat-like monsters of purely acoustic concreteness and purely intellectual abstraction. "Either we shall grow mute for a century [*na vek*], or philology [*slovesnost'*] will become for us a truly 'hermetic' cult, and the gift of explication (*hermeneuein*) will in a new way unite for us glossology with the gifts of spiritual edification—in concrete rationality."

The twin path of separation and reunion as the cure for the ailing word, indicated by Belyi, became the theme of Mandel'shtam's hermetic twin poem about the hayloft ("Ia ne znaiu, s kakikh por" and "Ia po lesenke pristavnoi," 1922). The immediate impetus for the writing of the poems was apparently Mandel'shtam's reading of Pasternak's *My Sister Life*,[11] which he seemed to consider as a species of trans-sense just as Khodasevich identified as trans-sense Mandel'shtam's own *Tristia*.[12] It has been pointed out by E. A. Toddes that Mandel'shtam used the futurist-formalist term *zaum'* nontrivially.[13] Indeed, he had expanded it to cover the disruption of the automatic nexus between the syntactic and the logical structure of the sentence, between the coherence of the poetic discourse and the message of the poetic utterance, and finally to account for the semantic emancipation of the poetry of the past from its historical referentiality as a prerequisite of its resurgence at the plane of literary synchrony. The latter notion was a salient feature of his model of literary history, as the following passage in his essay "Storm and Stress" clearly shows: "An intense involvement with the entire Russian poetry as a total, from the powerfully ungainly Derzhavin to the Aeschylus of Russian iambic verse—Tiutchev, had preceded futurism. At that time, approximately before the beginning of the World War, all the old poets suddenly seemed new. A fever of reappraisal and of hurried amends for historical injustice and short memory took hold of everybody. To the new inquisitiveness and the renovated hearing of the reader, the entire Russian poetry had then virtually presented itself as trans-rational."

As he returned in "The Hayloft" to the image of Verlaine's *chanson grise,* which he had already used in his essay "The Word and Culture" to describe the underlying naiveness of modern poetry, Mandel'shtam introduced a historical corrective in Belyi's identification of the "natural word" with the "futurism of language":

Ia ne znaiu, s kakikh por
Èta pesenka nachalas' . . .

I do not know since when
this ditty has begun . . .

"This ditty" for Mandel'shtam is as old as poetry itself. It has not begun with futurism and would not end with it. As for the "purloined [*uvorovannaia*] bond" between the logical life blood of poetry (vascular sclerosis is Mandel'shtam's metaphor for worn-out syntax in "Storm and Stress") and the metalogical "ringing [*zvon*] of the "dry grasses," it is rediscovered again "after a century" [*cherez vek*] (compare Belyi's *na vek,* quoted earlier) or whenever the poet invokes the purely phatic function of his art:

Ia khotel by ni o chem
Eshche raz pogovorit' . . .

About nothing I should like
One more time to have a talk . . .

An alternative to the union of "blood" and "the ringing of the withered grass" that is offered in the other "Hayloft" poem is the ultimate separation of sound and meaning, and the return of poetry and the poet to the "native sound scale" [*rodnoi zvukoriad*]. Here Mandel'shtam recapitulates his earlier idea of the synthesis between Verlaine's *musique avant toute chose* and Tiutchev's resigned denial of any possibility of genuine communication ("A thought uttered is a lie") in an image of "primeval muteness," the unborn pure tone that retains the pre-cosmic unity of being:

Ostan'sia penoi, Afrodita,
I slovo v muzyku vernis',
I serdtse serdtsa ustydis',
S pervoosnovoi zhizni slito.

Remain foam, Aphrodite;
And, word, revert to music;
And, heart, shy away from another heart,
Fused with the first principle of life.

("Silentium," 1910)

It is in the second "Hayloft," the one that begins with the lines "Ia po lesenke pristavnoi / Lez na vsklochennyi senoval" [I climbed the ladder leaned against / The disheveled hayloft], that the image of the shaggy fleece [*kosmatoe runo*] embodies the bestial nature (in Belyi's sense) of poetry in the "burning ranks" of what appears to be a primitively ecstatic collective ritual of hay-making (such a reading is supported by Mandel'shtam's reference, in a very free adaptation of Max Barthel's poem, to "our burning choral dance" and "the burning drunken epos that we are").

The new harmony, associated with the "trans-sense dream" [*zaumnyi son*], is shaggy, as is the world of the hayloft poem, and the song strokes the world the wrong way: "Protiv shersti mira poem, / Liru stroim, slovno speshim / Obrasti kosmatym runom."[14] The paronomastic potential of the word *kosmatyi* 'shaggy,' suppressed in these lines by means of a synonymic substitution[15] (*mir* 'world' instead of *kosmos*), had previously been fully and grotesquely realized by Belyi as he rhymed *kosmy* and *makrokosmy* in *Pervoe svidanie* [*First Meeting*, 1921].

The direct source of the hay images in Mandel'shtam's twin poem, the relevant lines of "The Steppe" in *My Sister Life*, has been described by K. F. Taranovsky (see n. 11). To this subtext one might add "Imelos'" ("Besides, there was a hayloft at hand"), "Dushnaia noch'" ("Sultry Night," with the image of "the grasses in the sack of thunderstorm" which invites comparison with Mandel'shtam's "sultry haystack" and the "sack with caraway"), and especially "Raspad" and its vision of the burning revolutionary pile of hay: "Ona, tumannaia, vzvilas' / Revoliutsionnoiu kopnoi," with a contrastive paronomastic implication of *kupina* 'burning bush' in this context of the apocalyptic 'world conflagration' (the image of the burning stack or pile of hay would later reappear in the "Star of Nativity" poem of *Doctor Zhivago*).

There is, however, an iconographic subtext which provides a key to Mandel'shtam's transformation of Pasternak's imagery into a hermetic symbol of trans-sense as the central principle of the universe, its core of densely packed chaos, as it were. The stanza that points directly at this subtext was not published by Mandel'shtam, who restored it from the original drafts as late as 1936:[16]

Raspriazhennyi ogromnyi voz
Poperek vselennoi torchit.
Senovala drevnii khaos
Zashchekochet, zaporoshit.

An unharnessed enormous wain
Sticks across the universe.
The ancient chaos of the hayloft
Will tickle and powder with dust.

This appears to be a metapoetic interpretation of the iconography of "The Hay Wain" by Hieronymus Bosch, a ladder leaning against it ("Ia po lesenke pristavnoi / Lez na vsklochennyi senoval"), the whole brutalized humanity gathering around it, a group of aesthetes on top, and the grieving Logos (Belyi's *Imia Rek*) in the cloud above.

In view of the meaning of the word *sennyi* attested in V. I. Dal"s *Slovar'*, *inoskazatel'nyi, proobrazovatel'nyi* [allegorical, prefigurative], *sennomudryi chelovek*, "one who understands the language of allegories," Mandel'shtam's choice of this baffling painting as a reflexive figurative representation of trans-sense begging to be explained is especially apt: several generations of Bosch scholars have been looking for the needle of demonstrable allegorical sense in this haystack.[17]

Allusions to "The Hay Wain" can be found also in the poetry of some of Mandel'shtam's contemporaries and friends. Kliuev in his chiliastic poem "Trud" ["Labor," 1918] seems to use the image in a more literal sense—with an implied affirmation of the superiority of peasant poetry: "Svit' sennyi voz mudree, chem sozdat' / 'Voinu i mir' il' Shillera balladu" [To put together a hay wain takes more ingenuity than to create / A "War and Peace" or a Schiller ballad]. Benedikt Livshits must have been aware of Bosch's painting as a subtext of "The Hayloft." In his long poem "Aeschylus"[18] he added a touch of his own to the solution of the haystack's riddle by partially paraphrasing Byron's epigram "The world is a bundle of hay" (the paraphrase is ambivalent because, in view of the double meaning of the word *mir* in modern spelling, 'world' or 'peace,' it may also contain a learned reference to the burning stacks announcing the fall of Troy in *Agamemnon*):

> Do you think the world is a heap of thundering hay,
> The embrasures of Pergamus, and the blood-swollen rivers?
> And the only light in your window is Helen . . .

These words of reproach are apparently addressed to Mandel'shtam the author of "The Hayloft" and of the "Apollonian" lines "if it were not for Helen, what would Troy alone be worth?" They contain a wistful evocation of the avant-garde's furious world and the futurist struggle against classicism in the name of the "pre-Homeric," "Pelasgian" harmony:

> Ty dumaesh', mir – èto vorokh gremiashchego sena,
> Boinitsy Pergama i krov'iu nabukhshie reki?
> I tol'ko i sveta v okne u tebia, chto Elena . . .
> O moiry, kakaia ustalost' smezhaet mne veki!
>
> Kuda ubezhat' ot muchitel'no iasnogo mira,
> Gde ne v chem tonut' moemu nenasytnomu vzgliadu,
> Gde lad pelazgiiskii utratila drevniaia lira
> I vkhodit, kak v larets, velikii Olimp v Iliadu?

The poem was written in 1933, when, on the eve of the new, forcibly introduced age of insipid pseudoclassicism, the muse of trans-sense was flying away from the Russian poetry.

It found an unexpected refuge, albeit constructively limited by appropriate thematic motivations, in the mature polyglot art of Vladimir Nabokov, enriched by the discoveries of modern poetry and unprecedentedly original in combining the full-blooded effect of vivid verisimilitude with a clearly expressed set toward "artistic device" as such.

In the twenties, Nabokov had been an astute reader of the formalists. His early short story "A Guide to Berlin" contains thoughtful allusions to Viktor Shklovsky's theoretical essays, as well as to *Zoo*. Similar to Tynianov, he radically transformed the biographical genre under the influence of Georgii Blok's admirable and unjustly forgotten book *Rozhdenie poeta. Povest' o molodosti Feta* [*The Birth of a Poet. A Tale of Fet's Youth*, 1924].

Some thought-provoking observations on trans-sense occurred already in Nabokov's earlier Russian prose. Psychologically and linguistically convincing specimens of trans-sense collocations are abundantly scattered all over his English-language books; their first appearance in his later Russian novels, *Invitation to a Beheading* and the unfinished *Solus Rex*, may have been inspired or provoked by Joyce's *Work in Progress*. An admirer of *Alice in Wonderland*, which he had translated into Russian, and of *Ulysses*, Nabokov was very critical of *Finnegans Wake*,[19] but often used it as a source book of multilingual puns and as a contrastive subtext to show how polyglottism and linguistic fantasy can better achieve an artistic purpose. The fictional languages in *Solus Rex* and in the English novels *Bend Sinister, Pale Fire,* and *Ada,* as well as certain personal and place names in *Lolita* and *Look at the Harlequins* offer numerous examples of obvious response to Joyce's macaronic experiments.

In *Pale Fire,* the Zemblan language, somewhat similar to the Russo-Norvegian trade dialect Moyapatvoya as well as to the imaginary pseudo-Scandinavian tongue of Thule in *Solus Rex,* is the invention of the novel's protagonist, a demented philologist. The trans-sense language in *Bend Sinister,* a lexical and grammatical hybrid of, mainly, Slavic and Germanic, is motivated, as in *Finnegans Wake,* by the subliminal paronomastic logic of a nightmare. The language consists in part of the phonetically transcribed or transliterated garbled fragments in a foreign language (e.g., dismembered quotations from Mallarmé's "L'Après-Midi d'un Faun,"[20] a symptomatic choice in view of the role fauns play in the Russian images of *zaum'*), of correctly rendered locutions and entire passages in Russian and in German (which might be perceived by an English-language reader who does not know these languages as trans-sense), and of punning charade-like multilingual portmanteau words and sentences, e.g.: *Domusta barbarn kapusta,* a local proverb translated as 'The ugliest wives are the truest' (*domusta,* from Eng. 'homely'; *barbarn,* a pseudo-Germanic plural of *barbara,* from Rus. *baba* 'married peasant woman'; *kapusta* [Rus. 'cabbage'], from the Latin formula *casta* et *pu*dica).

Thus Nabokov's trans-sense reverts to its original roots in barbarisms, exotic and incomprehensible foreign words and names, which nourished the glossolalia of Russian religious sects ("Daranata shantra / Sunkara purusha / Moia deva Lusha"), recorded by Mel'nikov-Pecherskii and discussed in terms of *zaum'* by Shklovsky,[21] and Khlebnikov's famous "Bobèobi pelis' guby," inspired, as Kornei Chukovskii pointed out,[22] by the hypnotic music of Indian names in Bunin's translation of *The Song of Hiawatha* ("Shli Choktosy i Komanchi, / Shli Shoshony i Omogi," and so on).

In his Russian novels, Nabokov examines two types of trans-sense. One is the trans-sense of the unfamiliar sound-shapes which may lose their poetic value once their objective referential meaning is ascertained (compare the use of the word *carminative* by the young poet ignorant of its meaning in Huxley's *Crome Yellow*). Here Nabokov's observations converge to a remarkable degree with those of the members of *Opoiaz,* as the following instance shows.

In *The Gift*, the poet and critic Koncheev, an imaginary interlocutor of the protagonist, Godunov-Cherdyntsev, makes the following confession: "When I was small, before sleep I used to say a long and obscure prayer which my dead mother [. . .] had taught me [. . .]. I remembered this prayer and kept saying it for years, almost until adolescence, but one day I probed its sense, understood all the words—and as soon as I understood I immediately forgot it, as if I had broken an unrestorable spell. It seems to me that the same thing might happen to my poems—that if I try to rationalize them I shall instantly lose my ability to write them."[23]

B. M. Eikhenbaum, in his early essay "O khudozhestvennom slove" ["On the Word in Art"] which remained unpublished until 1987, quoted and analyzed the following literary evidence: "V. G. Korolenko tells in *The Story of My Contemporary* how he loved to repeat 'Otche nash' ['Our Father'], understanding nothing and garbling the words; when his father explained to him the meaning of all the words and phrases the prayer lost for the child all of its enchantment."[24]

The other type of trans-sense considered by Nabokov represents the extreme limit of verbal reiteration 'by rote' (*tverzhenie*), the transformation of the word repeated into purely phonetical repetend,[25] which perhaps is pertinent to the Jakobsonian "weakening of referentiality in euphonic concentrations."[25a] This type has been described by L. P. Iakubinskii in his well-known article "On the Sounds of Poetic Language" (1919) on the basis of the material from William James[26] and from Mikhail Kuzmin's novel *Nezhnyi Iosif*: "James describes this phenomenon in the following way: 'if we look at an isolated printed word and repeat it long enough, it ends by assuming an entirely unnatural aspect. Let the reader try this with any word on this page. He will soon begin to wonder if it can possibly be the word he has been using all his life with that meaning. [. . .] It is reduced, by this new way of attending to it, to its sensational nudity.' The phenomenon of the word's 'nudity' is quite widespread, and everyone has probably observed it. [. . .] 'Roma? good; round like a dome. [. . .] Joseph began to repeat: Roma, Roma,—until the sounds lost for him their meaning and something enormous as the sky or a church's dome infused his soul."[27]

Nabokov devoted to this species of trans-sense a particularly revealing metaphysical digression in *The Gift*: "[. . .] pishesh', pishesh' adres, mnozhestvo raz, mashinal'no i pravil'no, a potom vdrug spokhvatish'sia, posmotrish' na nego soznatel'no, i vidish', chto ne uveren v nem, chto on neznakomyi,—ochen' stranno. . . . Znaesh': potolok, pa-ta-lok, pas ta loque, patolog,—i tak dalee,—poka 'potolok' ne stanovitsia sovershenno chuzhim i odichalym, kak 'lokotop' ili 'pokotol.' Ia dumaiu, chto *kogda-ni-*

bud' so vsei zhizn'iu tak budet" ["one writes an address heaps of times, automatically and correctly, and then all of a sudden one hesitates, one looks at it consciously, and one sees you're not sure of it, it seems unfamiliar—very queer. . . . You know, like taking a simple word, say ceiling and seeing it as 'sealing' or 'sea-ling' until it becomes completely strange and feral, something like 'iceling' or 'inglice.' I think that *some day* that will happen to the whole of life."][28]

Elsewhere in *The Gift* Nabokov developed this thought further as he described the awakening of his protagonist after an ecstatically happy dream about the reunion with his dead father: "At first the superposition of a thingummy on a thingumbob and the pale, palpitating stripe that went upward were utterly incomprehensible like words in a forgotten language or the parts of a dismantled engine—and this senseless tangle sent a shiver of panic running through him: I have woken up in the grave, on the moon, in the dungeon of dingy non-being. But something in his brain turned, his thought settled and hastened to paint over the truth—and he realized that he was looking at the curtain of a half-open window, at a table in front of the window: such is the treaty with reason—the theatre of earthly habit, the livery of temporary substance."[29]

Nabokov appears to recapitulate in these excerpts Shklovsky's philosophy of de-automatization in a reversed order of reasoning, proceeding from words to "things," rather than from "things" to words as Shklovsky did in his interpretation of a passage from Tolstoy's diaries ("Art as a Device," 1917),[30] but affirms together with Shklovsky, and contrary to the protagonist of Olesha's celebrated "Liompa," that the triumph of trans-sense over the "theatre of earthly habit" "some day" would be a transcendence rather than annihilation of meaning.

Notes

1. **Trans-sense:** Prince D. S. Mirsky, *Contemporary Russian Literature. 1881–1925*, N.Y., 1926 (repr. 1972), pp. 269, 274. **Transrational Language:** V. Markov, *The Longer Poems of Velimir Khlebnikov*, Berkeley, Los Angeles, 1962; *Russian Futurism: A History*, Berkeley, Los Angeles, 1968; *Modern Russian Poetry. An Anthology*, Indianapolis, N.Y., 1967, p. lxiii; K. Pomorska, *Russian Formalist Theory and its Poetic Ambiance*, The Hague, Paris, 1968, p. 83. **Metalogical Language:** D. Obolensky, "Introduction," *The Penguin Book of Russian Verse*, Harmondsworth, 1962, p. xlvii (also translation of Khodasevich on p. 307). **Metaintellectual Poetic Language:** R. P., "Futurism," *Dictionary of World Literature*, Ed. by J. T. Shipley, Totowa, 1966, p. 190. Roman Jakobson preferred, in his later years, to translate *zaumnaia poèziia* as **supraconscious poetry:** "Aljagrov's Letters" (1979), in *Selected Writings*, vol. VII, Berlin, N.Y., Amsterdam, 1985, p. 357. Compare **super-rational poetry, super-reason** for *zaumnaia poèziia* and *zaum'*, resp, in: Leon Trotsky, *Literature and Revolution*, transl. by Rose Strunsky, N.Y., 1925 (repr. Ann Arbor Paperbacks, 1960), pp. 133,

134. Paul Schmidt has ingeniously suggested **beyonsense** for *zaum'*, otherwise referring to *zaumnyi iazyk* as the **supra-rational, transcendental language of the future:** "Introduction," Velimir Khlebnikov, *The King of Time*, Tr. by P. Schmidt, Ed. by Ch. Douglas, Cambridge, Mass., and London, 1985, p. 3. **Transmental Language**: C. Barnes, *Boris Pasternak*, Vol. I, Cambridge, 1989, p. 163.
2. G. O. Vinokur, *Filologicheskie issledovaniia. Lingvistika i poètika*, Moscow, 1990, p. 284.
3. "Deklaratsiia zaumnogo iazyka." Reprinted in Kruchenykh's later editions and, more recently, in: V. Markov, Ed., *Manifesty i programmy russkikh futuristov*, Munchen, 1967, pp. 179–180. In the extensive bibliographic list appended to the edition of Vinokur, M. I. Shapir includes Kruchenykh's *Zaum'* (Baku 1921).
4. *The Gift*, London, 1963, p. 34; *Dar*, N.Y., 1956, p. 35. The 'subliminal labyrinth' in this passage should be compared with Nabokov's earlier, derogatory rendering of *zaum'* as 'submental grunt' (compare Maiakovsky's collection *Prostoe kak mychanie* [*As Simple as Mooing*, 1916]), the invention of the futurist poet Alexis Pan in *The Real Life of Sebastian Knight*, Norfolk, Conn., 1959, p. 28. Both the surname and the name of the poet are significant (Aleksandr Kruchenykh changed his name to Aleksei), and both were borrowed, subliminally, in all probability, from Aleksei N. Tolstoy's amusing picaresque novel *Ibikus*, in which Aleshka Pan and Fed'ka Arap are Odessa gangsters, "the local celebrities" (*Russkii sovremennik*, 1924, no. 3, p. 29).
5. On the excessively general use of the term *zaum'* ("a kind of surname, which is shared by all sorts of relatives and even namesakes"), see: Iu. Tynianov, *Arkhaisty i novatory*, Leningrad, 1929, p. 581.
6. R. Jakobson, "Noveishaia russkaia poèziia," *Raboty po poètike*, Moscow, 1987, p. 313.
7. On the futurists' progress from *Carrion Moon* to *Victory over the Sun*, see: A. Kruchenykh, "Ob opere 'Pobeda nad solntsem'," *Vstrechi s proshlym*, 7, Moscow, 1990, p. 511.
8. V. Khlebnikov, "Tezisy k vystupleniiu," *Sobranie proizvedenii*, V, Leningrad, 1933, p. 259.
9. A. Belyi, "Zhezl Aarona (O slove v poèzii)," *Skify*, Sbornik l, St. Petersburg, 1917, pp. 155–212.
10. The reference is to Pushkin's line *Parki bab'e lepetan'e* 'the old woman's babble of Parca' from "Verses Composed at Night during Insomnia," a seminal text for the Russian symbolism.
11. See: O. Ronen, *An Approach to Mandel'štam*, Jerusalem, 1983, p. 70, n. 10; K. Taranovsky, "Two Notes on Mandel'štam's 'Hayloft' Poems," *Text and Context; Essays to Honor Nils Ake Nilsson*, P. A. Jensen et al., Eds. Stockholm, 1987, pp. 122–127.
12. V. Kh., "O Mandel'shtam. 'Tristia.' Stikhi. Berlin, 1922," *Dni*, No. 13, Berlin, 12 November 1922. Reprinted in: V. Khodasevich, *Koleblemyi trenozhnik. Izbrannoe*, Moscow, 1991, pp. 518–519.
13. E. A. Toddes, "Mandel'shtam i opoiazovskaia filologiia," *Tynianovskii sbornik. Vtorye Tynianovskie chteniia*, Riga, 1986, p. 90.
14. On the relevance of Mandel'shtam's recurrent image of the "literary man's fur-coat" to this instance, see: K. Taranovsky, *Essays on Mandel'štam*, Cambridge, Mass., 1976, pp. 44–45. On I. Èrenburg's "Razdum'ia" (1921) [Liru razbil o kamen' severa / *Kosmatym runom obros.* / Na razvalinakh mira molchi, / Pushkina poldnevnaia tsevnitsa! / Varvar smeetsia, zabytyi mladenets krichit, / B'et krylami vspugannaia ptitsa] as an immediate subtext: O. Ronen, *op. cit.*, p. 70.
15. On dissimilation by means of synonymic substitution, see: O. Ronen, "Dva poliusa paronomazii," *Russian Verse Theory*, Ed. by B. Scherr and D. S. Worth, Columbus, Ohio, 1989, pp. 290–294.
16. Osip Mandel'shtam, *Stikhotvoreniia*, Leningrad, 1973, p. 130 and N. I. Khardzhiev's note on p. 282. On Mandel'shtam's practice of omitting those stanzas which disclose the thematic origin of the poem, see: O. Ronen, "Opushchennye strofy i podtekst. K istorii akmeisticheskikh tekstov," *Slavica Hierosolymitana*, III, 1978, pp. 68–76; M. L. Gasparov,

"'Zato što ruke tvoje . . .' Pjesma s odbačenim ključem," *Književna smotra*, Nos. 57–58, Zagreb, 1985, pp. 83–88.

17. See the summary of this painting's iconographic interpretations, in: R. H. Marijnissen, R. Ruyffelaere, *Hieronymus Bosch. The Complete Works*, Tabard Press, 1987, pp. 52–59.
    The relevance of "The Hay Wain" to Mandel'shtam's "Hayloft" has been noticed by R. D. Timenchik (oral communication).
18. Benedikt Livshits, *Polutoraglazyi strelets. Stikhotvoreniia. Perevody. Vospominaniia*, Ed. by P. M. Nerler and A. E. Parnis, Leningrad, 1989, p. 128 (first publication).
19. See: V. Nabokov, *Selected Letters 1940–1977*, San Diego, N.Y., and London, 1989, p. 350; B. Boyd, *Vladimir Nabokov. The Russian Years*, Princeton, 1990, p. 425.
20. The source was identified by Nabokov himself in his introduction to a later edition of *Bend Sinister* (Time Life Books, N.Y., 1964, pp. xvi–xvii).
21. V. Shklovskii, "O poèzii i zaumnom iazyke," *Sborniki po teorii poèticheskogo iazyka*, I, Petroograd, 1916, pp. 1–15. Repr. in: *Gamburgskii schet*, Moscow, 1990, pp. 45–58.
22. K. Chukovskii, "Ègofuturisty i kubofuturisty. Obraztsy futuristicheskikh proizvedenii," *Literaturno-khudozhestvennyi al'manakh izd-va "Shipovnik,"* no. 22, St. Petersburg, 1914, p. 144. Reprinted: K. Chukovskii, *Futuristy*, Petrograd, 1922; *Sobranie sochinenii*, vol. 6, Moscow, 1969, pp. 245–246. Compare: S. Bobrov, "Zvukovaia poèziia," *Literaturnaia èntsiklopediia. Slovar' literaturnykh terminov*, vol. I, Moscow and Leningrad, 1925, cols. 268–269.
23. *The Gift*, London, 1963, p. 322.
24. B. Eikhenbaum, *O literature*, Ed. by E. A. Toddes, M. O. Chudakova, and A. P. Chudakov, Moscow, 1987, pp. 333–334, n. on p. 493.
25. On monotonous repetition in Dada, see: V. Admoni, *Poètika i deistvitel'nost'*, Leningrad, 1975, pp. 185–188.

25a. R. Jakobson, *Raboty po poètike*, Moscow, 1987, p. 311.

26. W. James, *Psychology*, N.Y., 1900, p. 314. Iakubinskii refers to the noticeably altered Russian translation: V. Dzhems, *Psikhologiia*, St. Petersburg, 1911, p. 269. James' original text has been restored in our translation of the excerpt from Iakubinskii's paper.
27. L. P. Iakubinskii, "O zvukakh stikhotvornogo iazyka," *Sborniki po teorii poèticheskogo iazyka*, I, 1916, pp. 6–30. Reprinted in: L. P. Iakubinskii, *Izbrannye raboty. Iazyk i ego funktsionirovanie*, Moscow, 1986, p. 169.
28. *Dar*, N.Y., 1956, pp. 391–392; *The Gift*, London, 1963, p. 330.
29. *The Gift*, London, 1963, p. 336.
30. V. Shklovskii, "Iskusstvo kak priem," *Sborniki po teorii poèticheskogo iazyka*, II, Petrograd, 1917, pp. 3–14. Current reprint: V. Shklovskii, *Gamburgskii schet*, Moscow, 1990, pp. 58–72.

SLAVIC FOLKLORE AND LITERATURE

Russian Calendar Prose: The Yuletide Story[1]

YELENA DUSHECHKINA

St. Petersburg University

THE DECEMBER 1826 issue of *Moskovskii telegraf* (*Moscow Telegraph*) included a work entitled "Sviatochnye rasskazy" ("Yuletide Stories") from the pen of the journal's publisher Nikolai Polevoi. The narrative begins with the author's reminiscences of his longtime friends, old men from around Moscow, who "were living out and talking out their years" ("dozhivali i dogovarivali svoi vek") in various corners of the old capital. These oldsters had something to recount, not only to each other but to those young people who might prefer a quiet conversation by the fireplace to society balls and routs. And so now, this Yuletide, when Polevoi is writing his reminiscences, he recalls an old holiday evening which he spent with the old men of whom he was so fond. Their host on that evening gave the conversation a "needed" turn, since he had a "passionate custom always to speak of that which is appropriate to time and circumstance" ("strastnaia privychka govorit' vsegda o tom, chto prilichno vremeni i obstoiatel'stvam"). Since the conversation took place during Yuletide, one had to speak of nothing else but this holiday, he considered. Thus, he skillfully directed the conversation until the gathered friends started sharing their memories of happy holiday evenings of their youth. At the end, one of them, Shumilov, who had "traveled through half of Russia, and remembered all he had ever seen" ("ob"ezdivshii pol-Rossii, pomnivshii vse, chto kogda-libo videl"), told a Yuletide "horror story" ("strashnaia povest'") drawn from the history of medieval Novgorod. At the conclusion of his narrative, Polevoi promises readers to share with them many other such stories which he had heard from his old friends (Polevoi [pseud. N. P.] 1826).

Elementa, 1993
Vol. 1, pp. 59–74
Reprints available directly from the publisher
Photocopying permitted by licence only

Apparently, it was this publication by Polevoi which introduced into Russian literary and cultural usage the term "Yuletide story," which became so popular a few decades later. In several important ways, his text reconstructs the natural conditions in which such narratives operate. "Yuletide Stories" appears in the twelfth issue of Polevoi's monthly, and so the readers received it *in time for Yuletide*. He recreates the illusion of a text being shaped precisely *during Yuletide* ("and so now, by the way, during Yuletide, listen to . . ." ["i vot teper', kstati, *na sviatkakh*, poslushaite . . ."]); he reproduces a *Yuletide evening* spent with his favorite oldsters ("Like today, it was Yuletide" ["Byli, kak teper', sviatki"]); and, finally, the old men themselves *recall Yuletide evenings* of their youth, while one of them recounts a *Yuletide story*. Polevoi's work is shot through with events during and accounts of the holiday. Here, the author showed himself as a verbal artist who had internalized certain popular notions, at the core of which lies a need to, on a particular date, reminisce about and recount events which had occurred on the same day in the past. Because of this, "Yuletide Stories" may be regarded as an example of *calendar literature*.

This category of texts has been handled unevenly in literary scholarship. Some calendar works, such as ritual songs that are sung during winter holidays (*podbliudnye* ["platter songs"], *koliadki* and *shchedrovki* [Christmas and New Year's carols]), songs that accompany the rite of greeting of spring (*vesnianki*), songs that are performed on Midsummer Eve (St. John's Eve songs, or *kupal'skie pesni*), and so on, have been thoroughly studied. Others, namely *prose* works connected with holidays, generally have been neglected. Only in the last few years both traditional and written calendar texts in prose have become the focus of scholarly attention.[2]

The purpose of traditional oral prose texts connected with holidays is to inform members of the younger generation about holiday rites and to teach them the rules of behavior that are proper for a particular holiday. Thus, some of these texts describe holiday customs meals games and so on. In other words, by telling these stories the older people instruct the young about the holiday scenario. Other texts tell about extraordinary events that happened during holidays in the past. These stories, as a rule narrated by eyewitnesses, are supposed to teach the young people to behave properly during holidays.

The holiday requires a special type of behavior, since it is a peculiar sort of time, connected with the sphere of the sacral (Toporov 1982). The holiday is dangerous for two major reasons. On the one hand, the activity of evil spirits (*nechistaia sila*) is increased during holidays. On the other hand, man himself is more vulnerable during this time. Since calendar holidays are pagan in their origin, they provoke sinful behavior (that is, sinful from the point of view of Christianity): singing, dancing, getting together for games and pranks, putting on disguises, and especially divination. Therefore, one must be particularly careful and know how to protect oneself from danger: utter magical words, such as *chur menia;* cross oneself; solve the riddles given by a demon or evil spirit; and so on.

In order to teach the young how to deal with dangerous holiday situations, old people (usually women) recounted a special kind of stories.[3] There is evidence that such texts would usually be narrated on holiday evenings.[4] These stories concerned people who had met evil supernatural beings during holidays in the past. In addition to the instructive function, these stories also affected the audience emotionally: they produced and maintained an atmosphere that was festive, merry, and scary at the same time. Lev Tolstoi, in episodes of *War and Peace* devoted to the description of Yuletide festivities, accurately recreates this mixture of mirth and fear that was so characteristic of the holiday mood.

Stories told during Yuletide are especially abundant.[5] The reasons for their popularity are clear: Yule, situated on the boundary of the old and new years, and connected with the cult of the dying and reborn sun, was regarded as the main holiday of the calendar cycle. It is celebrated during the days of winter solstice, and the festivities last for two weeks: it begins on December 25 and ends on January 6, thus coinciding with the time between Christmas and Epiphany.

Several features of this period deserve special emphasis:

(1) Yuletide was regarded as a time when evil spirits were most active. As the Russian ethnographer Sergei Maksimov points out in his classic work on Russian supernatural beings, *Nechistaia, nevedomaia i krestnaia sila*, during Yuletide "the innumerable crowds of petty devils come out of the underworld and freely roam the earth frightening the people" (Maksimov 1903: 321).[6]

(2) Yuletide, spanning time past and time future, was seen as a period when people felt an especially acute desire to know their fate. To learn what will happen to them, they performed special actions known as *gadanie* (divination). Information received through these actions came from evil beings, and thus divination itself was both dangerous and sinful. To make possible contact with evil beings, a person would strip himself or herself of objects believed to ward off the evil spirits: take off a cross, a belt, or an icon. The most dangerous times for divination were Saint Basil's Eve (that is, New Year's Eve) and the night before Epiphany.

(3) Yuletide was a time for parties (*vechorki*). At these parties the young people danced and played traditional games. In the past these games had magical significance. They included mock weddings and funerals. The games were believed to promote fertility of the earth and people and to increase future harvest. According to many witnesses, they were marked by extensive license, a relaxation of ordinary rules of conduct.[7] Other games include impersonating a horse, a bear, a goat, and other animals. The old people considered these games sinful.
(4) Yuletide was a time when young people looked for a mate. After Yuletide came the wedding season (*rozhdestvenskii miasoed*, the post-Christmas fast-free period).
(5) Dreams experienced during Yuletide were considered prophetic.[8]

Overall, Yuletide is a very intensive period of the year. It is filled with both good fortune and danger; during Yuletide, one can find either happiness or death.

Stories narrated at Yule evenings tell about critical situations in which a person encounters evil beings. If in this situation a person displays cowardice, ignorance, or recklessness, he or she pays with life or health or looses a member of his or her family. One such story tells of a girl who was at a party. While sitting by a window, she saw outside a young man whom she knew. She pressed her lips to the glass, and the young man kissed her. She immediately went mad because the man was in actuality a forest spirit (*leshii*).[9] The girl's mistake lay in being unable to recognize a forest spirit in a human body (*oboroten'*).[10]

Another story involves five girls who went to the bathhouse to inquire about their future. In folk tradition, the bathhouse is considered an unclean place where evil spirits of all sorts dwell. The girls met there the *sviatochnitsy*—female spirits who appear only during Yuletide. They are horrible, ugly, and hairy all over. They cannot speak, they can only hum and dance. When they see somebody, they begin to tear at his or her flesh. Naturally, the girls were very frightened and tried to flee. But the *sviatochnitsy* ran after them, tearing off pieces of flesh from their bodies. At this point one of the girls, who remembered that the *sviatochnitsy* liked beads, ripped off her necklace and threw the beads to the ground. The *sviatochnitsy* started to pick the beads up and forgot about the girls. The girls thus escaped safely and ran home. Clearly, the knowledge of appropriate rules ensures safety for a person entering a dangerous place during Yuletide.[11]

Stories about encountering an evil being generally are called *bylichka* (memorate) or *byval'shchina* (fabulate). A *bylichka* concentrates on a mere recollection of a supernatural occurrence, whereas a *byval'shchina* concerns itself with plot development. Virtually all stories told during Yuletide fall in these two categories.[12]

Yuletide *bylichki* and *byval'shchiny* instruct their audience about places where a meeting with an evil being is more likely to occur, about situations especially favored by evil beings, and about the evil beings themselves.

Evil beings can confront humans in a foreign space, that is, in a space that lies outside their household or village, such as highways (in particular, crossroads), woods, and cemetaries. However, more familiar places, situated within the familiar space of a village or a household, such as bathhouses, barns, uninhabited houses, attics, basements, and the like, also can be dangerous.

Among the situations that might involve encountering evil beings are evening parties (*vechorki*), trips, all sort of activity in dangerous places, and, especially, divination. A girl who sets out to learn about her fate (and it is girls who are most frequently engaged in this activity) wants to invoke an evil spirit in order to quiz it about her future. Such encounters often end in tragedy. Ivan Panaev, in an essay about Yuletide celebrations during his childhood, recalls the following story. Panaev writes that at Yuletide evenings it was customary to tell "a story about a young lady from the countryside (*derevenskaia baryshnia*) who wanted to see her prospective bridegroom (*suzhenjy*) in a mirror. She secretly prepared in a bathhouse all items necessary for the divination; she went there at midnight, all alone, looked in the mirror, but instead of her prospective bridegroom she saw herself in a coffin; she fell on the ground, and was found dead in the morning. Who told about what the young lady lad seen in the mirror," Panaev inquires, "if she was found dead? This simple question did not occur to anybody, but nobody doubted that this story was true."[13]

In terms of plots, *bylichki* are usually very simple, like the story given by Panaev. A *bylichka* about divination requires the following basic elements: direct confirmation that the events actually occurred (a reference to the participant in or witnesses to the events), the mode of divination, the information received through the divination, and confirmation that this was accurate. All these elements are found in a story that tells about the following mode of divination: the members of a family put their wooden spoons into a barrel with water, stir the water, and leave the spoons there for the night. The next morning they come to check the positions of the spoons: if one of them is separated from the rest, the person to whom it belongs will certainly die during the coming year. The story goes as follows: a peasant family put their spoons into a barrel, went to bed, and in the morning discovered that the spoon that belonged to a twelve-year-old girl floated separately. The mother later described her reaction: "I took a look and gasped. 'Why, Ol'ka,' I told my daughter, You will die without fail this year!'" The mother was right: the girl died before St. Nicholas' day.[14]

Byval'shchiny are more complex in terms of plots. One story tells of a girl who inquired about her future bridegroom. She used a mode of divination that is called "inviting a bridegroom for supper" ("priglashenie suzhenogo na uzhin"). The story goes as follows: "One Yuletide evening girls got together for divination in a shack that did not have any icons. One of the girls was very pretty and also fearless. She suggested taking their crosses off. When the girls did so, they felt great fear. But the pretty girl went ahead and began the divination. She put a bowl of pottage on the table and said: "My bridegroom, come to have supper with me" ("Suzhenyi-riazhenyi, prikhodi ko mne uzhinat'"). Immediately they heard bells jingling, and a sled drove up to the entrance of the shack. A handsome man came in. He took the girl by the hand and led her to his invisible secret village. He married her, and she lived there in prosperity, but missed her family very much and wanted to go back. Her husband did not want to let her go. Her grief made her sick. Once she had a dream, in which a voice said that she could be saved only by her godmother. She asked her husband to bring her godmother to her. He did so. The godmother recognized her goddaughter and said to her: 'You are so thin! And you do not wear a cross!' She took off her own cross and put it on the sick girl. The very same moment the girl's husband turned black, burst out laughing, and started clapping his hands. Everything around them disappeared, and the girl with her godmother found themselves in the middle of an empty field, ten *versts* from their village. In the morning they returned home."[15] While the plot of this story is not terribly sophisticated, it still consists of several steps.

In addition to serious and frightening Yuletide stories, Russian tradition also presents stories that are supposed to elicit laughter. They tell of young fellows who deceive girls engaged in divination, often by impersonating evil beings. One such story involves the daughters of a priest, who went to the barn for divination. They inquired about their future husbands. The answer was supposed to come in the following form: a girl turns her bare bottom to the barn's window; if a hand that slaps a girl on the bottom is in a mitten, she will marry a priest; if it is bare, she will marry a peasant. At the time of divination a peasant was working in the barn. He decided to fool the girls and to play the role of an evil spirit. He slapped one of the girls with his bare hand. He began to beat the other girl so violently that she screamed madly. Her scared companions grabbed her by her dress and rushed her home. After that all of them fell ill with fever.[16] V. Zinov'ev, a scholar from Siberia who published a marvelous collection of *bylichki*, as well as a study of this genre, follows a folkloristic tradition in labeling such stories "pseudo-*bylichki*," "quasi-*bylichki*" (1987: 398). Their function is to dispel slightly the intense atmosphere of fear that reigned during Yuletide.

The literary Yuletide story emerges from oral tradition, appearing in written form at the beginning of the eighteenth century. The first of such stories is "The Tale of Frol Skobeev," which is usually included among the so-called "tales of everyday life" (*bytovye povesti*) of the late seventeenth century, but which, in the view of most scholars today, actually was written down only in the Petrine epoch.[17] In 1785, there appeared a literary reworking of this tale that included a significant enhancement of its Yuletide component.[18] Since the Yuletide stories belong to the class of calendar texts, their most convenient sphere of existence in written form is the periodical press which by its nature is connected with the idea of calendar cycle.[19] It was Mikhail Chulkov—the publisher of the magazine *I to, i sio* (*Bits of This and That*, 1769), the first Russian ethnographer, and a collector of folk poetry—who introduced this genre into mainstream literature (publishing three Yuletide stories prior to the holiday season itself)[20] and who bequeathed this tradition to the mass periodicals of the nineteenth century.

As Russian periodicals began to multiply and to serve the mass reader, they began to publish an enormous number of all sorts of calendar texts. Some demographic processes taking place in nineteenth-century Russia contributed powerfully to the proliferation of such texts. The main consumers of mass periodical press were peasants who had moved to the cities.[21] They brought with them traditional customs and beliefs, including the habit to celebrate agrarian holidays. Calendar texts published in the periodical press helped them to compensate for the lost festivities and ceremonies. To satisfy this need the periodicals filled the pages of their holiday issues with appropriate holiday production: pictures, poems, essays, and stories about a holiday. As noted above, the most popular time to tell calendar texts in periodicals were published during Yuletide, with Easter occupying the second place in this regard.

Many of the Yuletide stories published in the periodical press were connected in form and theme with Yuletide *bylichki*. In becoming assimilated into written literature, these traditional texts underwent certain transformations. The most obvious difference between the *bylichka* and the literary Yuletide story is that the latter is much longer than its predecessor. There are several reasons for that.

(1) The traditional story is told in the same scary and exciting atmosphere of Yuletide evenings in which the events of the story had happened. The literary Yuletide story is removed from this atmosphere, since the reader is free to read it at any time. The audience of the traditional narrative is closer to the text and can participate in its construction with their remarks and emotional reaction. We may recall Tolstoi's description in *War and Peace* of the atmosphere surrounding the telling of such a work:

> "Ah! ah!" screamed Natasha, rolling her eyes with horror.
> "Yes? And how . . . did he speak?
> . . . .
> "Now, why frighten them?" said Pelageya Danilovna.
> (Tolstoi 1966: 581–82)

Listeners unconditionally believe in these stories, since they themselves know similar tales or may even have been involved in like events.

Written Yuletide stories are artificial; they unfold in an unnatural context. They only imitate the real Yuletide situation with a greater or lesser degree of success. Because of that authors often try to recreate this situation by beginning with a special exposition which presents a group of characters engaged in a conversation (*sviatochnaia beseda*). This conversation frequently turns to Yuletide subjects. One of the characters then narrates an appropriate story.

For example, in an 1896 story by the minor prose writer Kazimir Barantsevich, "Gusarskaia sablia" ("A Hussar's Sabre"), the narrator first describes in detail the setting of a Yuletide evening in a high society house in Saint Petersburg. He thus creates a holiday atmosphere and prepares the reader for a tale, which is then told by one of the characters, a colonel. Another example is found in a story by Kot-Murlyka (N. P. Vagner), "Liubka": "A long, long time ago, about forty-five years ago, or perhaps more, the four of us sat round the fireplace on Christmas Eve and listened to a story by the extraordinarily kind old man Petr Nefed'ich Trutkov" (1882). After this opening sentence comes Trutkov's own narrative.

(2) The traditional *bylichka* or *byval'shchina* does not need ethnographical explanations: its audience, as a rule, is familiar with the details of the holiday ritual. The presence of ethnographic inserts in many texts collected by folkloristic expeditions reflects only the fact that they were recorded in unnatural situations: told at the wrong time, and for the wrong audience. By comparison, consumers of written Yuletide stories are often removed from the holiday tradition. They need to be told the details of everyday life in the country and, especially, of the Yuletide ceremony. This explains why literary Yuletide stories contain a lot of ethnographical material. For some writers

the presentation of such material becomes the main reason to write a story. For example, the primary focus of a tale by Fedor Nefedov, "Chudnaia noch'" ("A Wondrous Night"), is the detailed description of a Yuletide night in a Russian village. This and similar stories are a rich source of information about Yuletide customs in different regions.[22] Significantly, writers of Yuletide and other prose calendar tales (N. Polevoi, N. Pogodin, N. Gogol', A. Marlinskii, and so on) played a significant role in developing the discipline of Russian ethnography during the 1820s–1830s.

(3) During the nineteenth century, written Yuletide stories are influenced by Russian psychological prose: authors try to describe their characters' emotions and explain their actions in detail. Where the narrator in a traditional Yuletide text limits himself to statements of fact, such as "The peasant was dumbstruck," "The girl became frightened and didn't know of what to talk with him," "A character in the story was very frightened" (Dushechkina, Ed. 1988c: 15, 22), the author of a literary tale describes a process: the emergence of a feeling of fear, its rise, and culmination. We can see that in Aleksandr Budishchev's story "Riazhenye" ("The Mummers"). In this 1886 work a character runs home through the woods on a Yuletide night. He becomes more and more frightened, and, ultimately, his fear leads him to mistake thieves in his house for devils. The author dwells on the description of his character's fear, from initial anxiety to the final paralyzing terror.

(4) Traditional *bylichki*, as a rule, lack an exposition: they begin with a description of the action itself. The narrator starts with such temporal references as "once," "one day," or "last year" and then turns to the events of the story. The authors of written Yuletide stories almost always provide large expositions. These not only serve to create a Yuletide atmosphere and to present ethnographical material, but also can unite several sub-plots. The composition of the literary Yuletide story is thus much more sophisticated. For example, in the story "Zamaskirovannyi" ("The Masked One"), by an anonymous author, the Yuletide episode is preceded by life stories of its main characters, a wife and a husband. The Yuletide story itself is used to introduce a surprise turn of the plot: one of the Yuletide guests turns out to be the wife's lover from the past, and she runs off with him (Nekto [pseud.] 1886).

As a result of all these changes the Yuletide episode in a literary Yuletide story can occupy only a small portion of its text. It is overshadowed by ethnographical and psychological details, as well as by elements serving compositional needs. Thus, in A. A. Shakhovskoi's story "Nechaiannaia svad'ba" ("The Unplanned Wedding") the Yuletide episode occurs at the very end of the narrative, actually forming its denouement (Shakhovskoi 1834).

But even more significant are the differences in the presentation of the Yuletide event itself. *Bylichki* present supernatural events as truthful: they are not concerned with explaining the mysterious and the supernatural. The Russian literary Yuletide story seldom leaves the supernatural without a scientific, psychological, and other realistic motivation. Plots similar to Gogol's "A Night Before Christmas" are rare in the Russian literary tradition.[23] The literary Yuletide tale sometimes starts with an assumption that the supernatural does not exist. Subsequently, the events in the story seemingly contradict this assumption. The protagonists, however, are harmed not because of their contact with the supernatural, but because they momentarily doubt their own conviction that the supernatural is not real. They fall ill or die from the resulting shock.[24] The author, however, does not share his characters' mistake and leads the reader to a realistic explanation of the events. The story thus acquires elements of a detective plot.

The collision between seemingly supernatural events and their realistic explanation sometimes can turn the story into a humorous one. Jocular Yuletide stories were a favorite in the late nineteenth century, they were published in large numbers in the burgeoning humoristic press (including such weeklies as *Razvlechenie* [*Amusement*], *Budil'nik* [*Alarm-clock*] and [*Oskolki* [*Fragments*]). The stories—some from the pen of N. Leikin and the young Chekhov—often tell about a meeting with an evil spirit. Then they show that this meeting was not real and that it can be explained by the character's alcoholic delirium, a condition connected with the Russian expression *napit'sia do chertikov,* to get so drunk that little devils appear before one's eyes, or, literally, "to get drunk to the little devils." In these works the supernatural element is used very liberally, since the realistic explanation can justify even the wildest fantasy. For example, in the story "Zaezzhii gost" ("The Guest Who Dropped In"), one of the narrators tells about the circumstances of his grandfather's death. His account is as follows: A stranger came to a Yuletide party. He turned out to be a devil. He made the narrator's grandfather drunk. Then he took the grandfather with him to buy more vodka. Their journey included such unlikely (from the point of view of the purpose of their journey) and dangerous places as woods, ravines, river banks, and cemeteries. Finally, the devil threw the grandfather into a pile of snow and left him there. After that the grandfather grew sickly and died within a month despite the fact that before this occurrence he had enjoyed good health. Did the devil really take the grandfather on this fatal journey, or had it happened only in the grandfather's drunken imagination? The author of the story leaves this question open (* * * [pseud.] 1875).

Thus the literary Yuletide story evolved not so much from the real *bylichka*, but rather from the pseudo-*bylichka*, which treats scary Yuletide events as a joke or a mistake by the characters in the tale.

This joke, however, often ends tragically. "Gadanie" ("Divination"), an 1880 story by Ivan Kupchinskii, tells of a girl who was smart and educated and did not believe in divination. Still, her servant persuaded her to look in a mirror to find out about her bridegroom who was supposed to be returning from the war, and whom she expected at any moment. While she was sitting in front of the mirror, her bridegroom at last arrived and decided to show himself to her in the mirror as her intended (*suzhenyi*). When the girl saw his reflection, she was suddenly filled with doubts about her lack of faith, and became so terrified that she died that very minute.

The literary texts described above certainly do not constitute great literature. They use banal narrative technique, a limited number of topics, and, frequently, a primitive language. Yet there are several reasons why the study of these stories holds the promise of significant results for both literary theory and literary history.

(a) Yuletide stories constitute a paradigm case of the assimilation of traditional genres into written literature. They belong to written literature, yet are still very much connected with the oral tradition. Thus, they form an intermediate stage between literary and traditional texts.

(b) The history of the Yuletide story represents a striking example of the rise, development, and decay of a generic tradition. The literary Yuletide tale, which arose on the basis of popular calendar stories, inherited from the latter both its system of protagonists and a limited set of plots involving supernatural phenomena. This well-defined set of narrative accessories becomes maximally enhanced by, on the one hand, plots based on a host of beliefs and trends fashionable in nineteenth century Russian society (mesmerism, spiritualism, telepathy, and so on), and, on the other, a new tradition of sentimental, moralizing Christmas texts that arose under the impact of Dickens (these include works by such writers as D. Grigorovich, F. Dostoevskii and N. Leskov).[25] These stories are examples of the almost unknown category of calendar prose, a special sort of literature that is consumed at certain times and has a special impact on the reader during these times. Writers, who easily "learned" the poetics of the Yuletide story, for the most part followed the path of least resistance, producing works which were but variations on a limited number of basic plots and themes. At the end of the nineteenth century, the inertia of the genre was only occasionally broken in works of unusual artistic merit (Leskov, Chekhov); more often, the lack of fresh spirit in the genre was reflected in the creation of *parodies*, which mocked both its narrative conventions and the context in which these were employed.

(c) Yuletide stories are part of mass literature in Russia, a phenomenon which only recently has become the object of intensive study. Yuletide texts and similar fictions used to make up the bulk of the habitual reading of the ordinary Russian reader, whose tastes, in turn were formed by such writings. Further documentation and analysis of such literature is needed if we are to grasp fully the psychology of the nineteenth- and early-twentieth-century russian mass reader—*literature* but still *uncultured.*

(d) Finally, during the Silver Age, the tradition of Yuletide and other calendar texts, in verse as well as prose, is unexpectedly resuscitated and becomes a stimulus and reservoir of artistic resources for leading Russian writers and poets, especially those associated with Symbolism and post-Symbolist movements.[26] As a result, such principal figures in twentieth-century Russian literature and culture as Alexander Block, Anna Akhmatova, Boris Pasternak, and, most recently, Joseph Brodsky create major works in which calendar thematics become interwoven with global historical and cultural meanings and associations.[27] We know the first-rate Russian literature rather well. However, our knowledge of Pushkin, Gogol', and especially such writers as Leskov and Chekhov, whose heritage includes first-rate examples of the Yuletide story, is incomplete without an understanding of their background, namely, mass literature. Mass literature, including the Yuletide story, played its modest but noticeable rule in Russian literary and cultural history.

Notes

1. I would like to thank Irina Reyfman (Columbia University) for her assistance in preparing this article.
2. Henryk Baran and myself almost simultaneously became interested in this genre: he began to study the late-nineteenth- and early-twentieth-century literary Christmas story, and I began to research traditional texts and literary Christmas stories of the eighteenth and nineteenth centuries. See Baran (1989; 1992a; 1992b), Dushechkina (1982; 1986; 1987; 1988a; 1988b, 1988c; 1989a; 1989b; 1990a; 1990b; 1991a; 1991b; 1992a; 1992b), Dushechkina and Baran (1992). Calendar literature has also recently been analyzed by Maya Kucherskaia in her student *diploma* thesis at Moscow University; see also her recent publication of a selection of Yuletide stories in *Druzhba narodov* 1 (1992): 222–270.

3. Extensive records of such special narratives are held in the archives of the Ethnographic Bureau of Prince Tenishev (Etnograficheskoe biuro kniazia V. I. Tenisheva) in the State Museum of Ethnography and Anthropology of the Peoples of the U.S.S.R. (Gosudarstvennyi Muzei Etnografii i Antropologii Narodov SSSR—henceforth GME), *fond* 7, *op.* 1. For a description of the holdings, see Pomerantseva (1971).
4. See, for example, Sakharov (1838: 64); Peizen (1903: 321); Dilaktorskii (1898: 134).
5. I use the term "Yuletide" instead of "Christmastide" in order to emphasize the pagan origin of the holiday.
6. See also Makarenko (1913: 48).
7. See the discussion in Preobrazhenskii (1864).
8. See, for example, the episode of Tat'iana's prophetic dream in *Eugene Onegin*. Also see the undermining of Yuletide expectations in V. Zhukovskii's ballad "Svetlana."
9. In this story, the spirit was called, respectfully, *Zvonkovskii*.
10. For the text of the story, see Novozhilov 1909.
11. GME, *fond* 7, *op.* 1, *d.*, 493, *l.* 31. Published in Dushechkina, Ed. 1988c: 25–26. On *sviatochnitsy*, see Pomerantseva (1975: 84). Popular belief also encompassed other "seasonal devils." See Makarenko (1913: 48), Tolstoi (1976: 302).
12. On such stories, see the discussion by Pomerantseva (1968).
13. Panaev's story, "Proshedshee i nastoiashchee (sviatki dvadtsat' piat' let nazad i teper')" ["The past and the present (Yuletide twenty-five years ago and today)], is published in Dushechkina, Ed. 1991a: 85–101.
14. See GME, *fond* 7, *op.* 1, *d.* 149, *l.* 22 (*ob.*). Published in Dushechkina, Ed. 1988c: 39.
15. GME, *fond* 7, *op.* 1, *d.* 493, *l.* 27. Published in Dushechkina, Ed. 1988c: 64.
16. GME, *fond* 7, *op.* 1, *d.* 141, *l.* 3–3 (*ob.*). Published in Dushechkina, Ed. 1988c: 76.
17. On the "Yuletide component" of "The Tale of Frol Skobeev," see Panchenko's contributions in *Istoriia 1* (1980: 380), *Istoriia 2* (1980: 377–415) and Dushechkina (1986).
18. See Novikov (1785). On it, see Dushechkina (1982).
19. See Dushechkina (1987), Baran (1992).
20. See *I to, i sio*, 1769, weeks 47 and 49. On Chulkov's Yuletide texts, see Dushechkina (1990a).
21. On this process, see Brooks (1985), as well as a recent study by Reitblat (1991).
22. Nefedov's story is included in his collection of Yuletide stories (1895: 185–214).
23. One example of a genuinely "fantastic" story is G. P. Danilevskii's "Progulka domovogo" ("The House-spirit's Stroll") (1901).
24. See, for example, a story by N. N. [pseud.] "Koldun—mertvets—ubiitsa" ("The sorcerer—the dead man—the killer," 1832), and the tale by M-in, "Rokovaia shutka" ("The fateful jest," 1883).
25. This tradition is described in greater detail by Baran (1992b).
26. On this revival, see Baran (1992b).
27. On Blok's use of Yuletide motifs, see, for example, Gasparov and Lotman (1975), Lotman (1981), and Merlin (1985). Brodsky's "Christmas" texts are discussed in an unpublished seminar paper by Ye. Rudneva (University at Albany).

Works Cited

Baran, H. 1989. "Paskha 1917 goda: Akhmatova i drugie v russkikh gazetakh." In: *Akhmatovskii sbornik I // Anna Akhmatova: Recueil d'articles*. Ed. Serge Deduline and Gabriel Superfin, 53–75. Paris: Institut d'Études slaves.

—.1992a. "Majakovskij's Holiday Poems in a Literary-Cultural Context." In: *Studies in Poetics: Festschrift for Krystyna Pomorska*. Ed. E. Semeka-Pankratov, 200–237. Columbus, Ohio: Slavica [forthcoming].

—.1992b. "The Tradition of Religious Holiday Literature and Russian Modernism." In: *Christianity and its Role in the Culture of the Eastern Slavs*, vol. 3, Eds. B. Gasparov, R. Hughes, et al. University of California Press [forthcoming].

Barantsevich, K. 1896. "Gusarskaia sablia. Rozhdestvenskii rasskaz" ("A Hussar's Saber. A Christmas Story"), *Vsemirnaia illiustratsiia*, no. 1456 (*Rozhdestvenskii nomer*), 671–676. Reprinted in Dushechkina, Ed. 1988c: 54–63.

Brooks, J. 1985. *When Russia Learned to Read: Literacy and Popular Literature, 1861–1917*. Princeton: Princeton University Press.

Budishchev, A. 1886. "Riazhenye. Sviatochnyi rasskaz." *Oskolki*, no. 52, 4. Reprinted in Dushechkina, Ed. 1888c: 16–21.

Danilevskii, G. P. 1901. "Progulka domovogo." In his: *Sochineniia*, vol. 19: *Sviatochnye rasskazy*, 50–54. Spb.

Dilaktorskii, protoier. 1898. "Sviatochnye shalosti v Pel'shemskoi volosti, Kadnikovskogo u. Vologodskoi gub.." *Etnograficheskoe obozrenie*, no. 4:133–135.

Dushechkina, E. V. 1982. "O kharaktere literaturnykh peredelok XVIII v. ('Novgorodskikh devushek sviatochnyi vecher' I. Novikova)." In: *Uchebnyi material po teorii literatury: literaturnyi protsess i razvitie russkoi literatury XVIII–XX vv.* Ed. A. F. Belousov. Tallin: Tallinskii pedagogicheskii institut.

—.1986. *Stilistika russkoi bytovoi povesti XVII v. (povest' o Frole Skobeeve). Uchebnyi material po drevnerusskoi literature.* Tallin: Tallinskii pedagogicheskii institut.

—.1987. "O kharaktere perezhivaniia vremeni v russkoi periodike XIX veka." In: *Prostranstvo i vremia v literature i iskusstve: teoreticheskie problemy, klassicheskaia literatura. Metodicheskie materialy po teorii literatury.* Eds. F. P. Fedorov et al., 38–40. Daugavpils: Daugavpilsskii pedagogicheskii institut.

—.1988a. "Chekhov i problema kalendarnoi prozy." In: *Literaturnyi protsess i problemy literaturnoi kul'tury. Materialy dlia obsuzhdeniia.* Ed. V. Neverdinova, 83–87. Tallin: Tallinskii pedagogicheskii institut.

—.1988b. "Noch' pered Rozhdestvom' i traditsiia russkogo sviatochnogo rasskaza." In: *Nasledie N. V. Gogolia i sovremennost'. Chast' perfaia. Tezisy dokladov i soobshchenii nauchno-prakticheskoi Gogolevskoi konferentsii (24–26 maia 1988 goda).* Eds. G. V. Samoilenko et al, 21–22. Nezhin: Nezhinskii gosudarstvennyi pedagogicheskii institut.

—,Ed. 1988c. *Russkaia kalendarnaia proza: antologiia sviatochnogo rasskaza. Uchebnye materialy po spetskursu.* Tallin: Tallinskii pedagogicheskii institut.

—.1989a. "Peterburgskii mif' i russkii sviatochnyi rasskaz." In: *Antsiferovskie chtenniia. Materialy i tezisy konferentsii (20–22 dekabria 1989 g.).* Comp. A. I. Dobkin and A. V. Kobak, 121–125. Leningrad: Leningradskoe otdelenie Sovetskogo fonda kul'tury.

—.1989b. "Ruskaia Kalendarnaia progz 20–30-kh godov XIX veka." In: *Istoriko-literaturnyi protsess. Metodologicheskie aspekty. Nauchno-informatsionnye soobshcheniia.* vol. 2. *Russkaia literatura XI-nachala XX vv..* Eds. L. S. Sidiakov et al., 32–34. Riga: Latviiskii GU.

—.1990a. "'Sviatochnye istorii' v zhurnale M. D. Chulkova 'I to i sio'. In: *Literatura i fol'klor: voprosy poetiki. Mezhvuzovskii sbornik nauchnykh trudov,* 12–22. Volograd: Volgogradskii gosudarstvennyi pedagogicheskii institut.

—.1990b. "Sviatochnyi rasskaz: vozniknovenie i upadok zhanra. "In: *Prostranstvo i vremia v literature i iskusstve. Metodicheskie materialy po teorii literatury.* Eds. F. P. Fedorov et al., 42–44. Daugavpils: Daugavpilsskii pedagogicheskii institut.

—.1991a. *Peterburgskii sviatochnyi rasskaz.* Leningrad: Petropol'.

—.1991b. "'Sviatochnaia' proza M. P. Pogodina." In: *Russkii romantizm: Prostranstvo i vremia.* Eds. F. P. Fedorov et al., 118–135. Daugavpils: Daugavpilsskii pedagogicheskii institut.

—.1992a. "N. S. Leskov i traditsiia russkogo sviatochnogo rasskaza." In: Collection of articles on nineteenth century Russian Literature. Spb.: Nauka (in press).

—1992b. *Sviatochnye rasskazy.* Moscow: Rudomino (in press).

Dushechkina, E., and H. Baran, Eds. 1992. *Chudo rozhdestvenskoi nochi: Sviatochnyi rasskaz.* Spb.: Khudozhestvennaia literatura (in press).

Gasparov, B. M., and Iu. M. Lotman. 1975. "Igrovye motivy v poeme 'Dvenadtsat". In: *Tezisy I Vsesoiuznoi (III) konferentsii "Tvorchestvo A. A. Bloka i russkaia kul'tura XX veka,"* 53–63. Tartu: Tartuskii GU.

Istoriia 1. 1980. *Istoriia russkoi literatury.* 4 vols. vol. 1. Leningrad: Nauka.

Istoriia 2. 1980. *Istoriia russkoi literatury X–XVII vekov.* Ed. D. S. Likhachev. Moscow: Prosveshchenie.

Kupchinskii, I. 1880. "Gadanie. Nedavnii sluchai." *Gazeta Gattsuka,* no. 1, 5 January, 10–15. Reprinted in Dushechkina, Ed. 1988c: 78–87.

Lotman, Iu. M. 1981. "Blok i narodnaia kul'tura goroda." In: *Blokovskii sbornik,* no. 4, 7–26. Tartu: Tartuskii GU.

Makarenko, A. A. 1913. *Sibirskii narodnyi kalendar' v etnograficheskom otnoshenii.* Spb.

Maksimov, S. V. 1903. *Nechistaia, nevedomaia i krestnaia sila.* Spb.

Merlin, V. V. 1985. "'Snezhnaia maska' i 'Dvenadtsat" (K voprosu o sviatochnykh motivakh v tvorchestve Bloka." In: *Uchenye zapiski Tartuskogo Gosudarstvennogo universiteta,* no. 680, 19–28. Tartu: Tartuskii GU.

M-in [pseud.] 1883. "Rokovaia shutka." *Novoe vremia,* no. 2812, 25 December, 4.

Nefedov, F. 1895. *Sviatochnye rasskazy.* Moscow.

Nekto [pseud.] 1886. "Zamaskirovannyi. Sviatochnyi rasskaz." *Birzhevye vedomosti,* no. 352, 25 December, 1. Reprinted in Dushechkina, Ed. 1988c: 69–75

N. N. [pseud.] 1832. "Koldun–mertvets–ubiitsa." *Molva,* no. 1, 2–4.

Novikov, I. 1785. "Novgorodskikh devushek sviatochnyi vecher, sygrannyi v Moskve svadebnym." In his: *Pokhozhdeniia Ivana gostinogo syna i drugie povesti i skaski.* Part 1, 112–159.

Novozhilov, A. 1909. "Derevenskie 'bisedy' (Khotenovskaia vol., Kirillovskii u., Novgorodskaia gub.)." *Zhivaia starina* 1:68.

Peizen, G. 1903. "Etnograficheskie ocherki Minusinskogo i Kanskogo okrugov Eniseiskoi gubernii." *Zhivaia starina* 3:297–357.

Polevoi, N. A. [pseud. N. P.] 1826. "Sviatochnye rasskazy." *Moskovskii telegraf* 12:103–123, 139–192.

Pomerantseva, E. V. 1968. "Zhanrovye osobennosti russkikh bylichek." In: *Istoriia kul'tury, fol'klor i etnografiia. VI Mezhdunarodnyi s"ezd slavistov,* 274–293. Moscow: Nauka.

—.1971. "Fol'klornye materialy 'etnograficheskogo biuro' V. I. Tenisheva." *Sovetskaia etnografiia* 7:137–147.

—.1975. *Mifologicheskie personazhi v russkom fol'klore.* Moscow: Nauka.

Pr<eobrazhen>skii, N. 1864. "Bania, igrishche, *slushan'e* i shestoe ianvaria." *Sovremennik,* vol. 10:499–522.

Reitblat, A. 1991. *Ot Bovy k Bal'montu: ocherki po istorii chteniia v Rossii vo vtoroi polovine XIX veka.* Moscow: Izd. MPI.

Sakharov, I. P. 1838. "Russkie sviatki." *Literaturnoe pribavlenie k Russkomu invalidu,* no. 4, 22 January.

Shakhovskoi, A. A. 1834. "Nechaiannaia svad'ba. Moskovskaia byl'. Rasskaz molodogo polkovnika K. . . . D . . . . ." *Biblioteka dlia chteniia,* vol. 2, 32–48. Reprinted in Dushechkina and Baran, Eds. 1992.

Tolstoi, L. 1966. *War and Peace.* Norton Critical Edition. Tr. Louise and Aylmer Maude. Ed. George Gibian. New York: W. W. Norton.

Tolstoi, N. I. 1976. "Iz zametok po slavianskoi demonologii. 2. Kakov oblik d'iavol'skii." *Narodnaia graviura i fol'klor v Rossii XVII–XIX vv.,* 288–319. Moscow,: Sovetskii khudozhnik.

Toporov, V. N. 1982. "Prazdnik." In: *Mify narodov mira,* vol. 2, 329–331. Moscow: Sovetskaia entsiklopediia.

Vagner, N. P. (pseud. Kot-Murlyka) 1882. "Liubka (Rozhdestvenskii rasskaz)." *Novoe vremia*, no. 2454, 28 December, 2. Reprinted in Dushechkina, Ed. 1988c: 89–99.

Zinov'ev, V. P. 1987. "Bylichka kak zhanr fol'klora i ee sovremennye sud'by." In: *Mifologicheskie rasskazy russkogo naseleniia Vostochnoi Sibiri*, 381–400. Novosibirsk: Nauka, Sibirskoe otdelenie.

* * * [pseud.]. 1875. "Zaezzhii gost'. Sviatochnyi rasskaz." *Razvlechenie* 51: 403–407.

Khlebnikov's Solar Myth Reexamined

HENRYK BARAN

University at Albany, SUNY

"The chief conclusions remind one to some extent of the fascinating debates among contemporary anthropologists on the character, extent, and range of application of the idea of myth. These conclusions can roughly be paraphrased in the following way: poetry and myth are two closely linked forces that are at the same time quite contradictory. The collision between these two elemental forces lies in the fact that poetry is oriented toward variation, while myth aims at invariance."

Jakobson and Pomorska, *Dialogues* (p. 150)[1]

IN 1914, VELIMIR KHLEBNIKOV published the first of his "supersagas" (*sverkhpovesti*), "Otter's Children" ("Deti Vydry"): a montage of diverse, largely autonomous texts written between 1911 and 1913 that serves as a recapitulation of major themes in his writings prior to World War I. The six-part "Otter's Children" begins with a recreation of a cosmogonic myth in which the protagonist, Son of Otter (*Syn Vydry*), playing the role of culture-hero, successfully confronts the three suns that shine in the sky of a primeval Earth and render inhabitable this original world:

... The first sun is white, the second smaller—ed surrounded by a ring of bluish light—and the third is black, with a green corolla. What seem to be words of complaint and anger in a strange language can be heard. At one corner of the curtain the tip of a wing is just visible. A winged spirit with a black spear in his hand appears above the golden shoreline, his eyes bright with ill-will. A spear shivers and flies, and the red sun falls as if it were setting, drops like a red pearl into the sea. The land changes appearance and begins to darken. A few shoots of green immediately spring up on the cliff. Floods of birds.

Standing on the dying sun, they raise their hands and sing a wordless hymn of praise to someone. Then Son of Otter—dark-haired, dark-skinned, his round head covered with curls—pulls out the spear and with a rush of black wings attacks the black sun, beating his wings against the air for support. And that sun too falls into the water. Deer and other animals appear.

The earth grows immediately darker. The sky becomes bright blue again. The sea changes

Elementa, 1993
Vol. 1, pp. 75–88
Reprints available directly from the publisher
Photocopying permitted by licence only

color—its black hue, gleaming with red, becomes green. Otter's Children grasp hands and for the first time set foot on earth. (CW II: 278-279)

The poet himself comments on the genesis of "Otter's Children" in his later, retrospective "Self-Statement" ("Svoiasi"). He notes first: "... using the legends of the Oroches—the oldest legends in the world—about the fiery fabric of the earth, I made Son of Otter attack the sun with a spear and destroy two of the three suns, the red one and the black one." A few lines down he claims that the "sagas of the Oroches, that ancient Amur tribe, had a profound influence on me, and I conceived the idea of creating a pan-Asian consciousness in my poems." (CW II: 6)

Two decades ago, I was able to establish the specific work from which the poet took the Oroche materials which he used to shape not only "Otter's Children" but also several other poetic and prose texts (Baran 1973). This proved to be the monograph report of an expedition investigating the culture of this minor Siberian people by the Society for the Study of the Amur Region (Margaritov 1888).[2] Subsequently, I discussed the function of the cosmogonic narrative in "Otter's Children," showing that it operates as a generator of the work as a whole: the heroism of the culture hero, incarnated in a variety of historical and legendary *dramatis personae*—including Khlebnikov himself—is replicated in many of the episodes that constitute the bulk of this ambitious if flawed text. Khlebnikov's heliomachy was set in the context of similar texts by other Futurists, especially A. Kruchenykh and V. Maiakovskii: I suggested that it may have been the source of the Hylaeans' focus on such thematics, and that a heroic myth lay at the root of the Hylaeans' self-definition as a poetic school (Baran 1976, 1983).

In this paper, I return to the subject of Khlebnikov's use of the episode depicting a hero's conflict with the sun. My focus is on two related questions: (1) given the wealth of mythological traditions, and Khlebnikov's own knowledge of so many of them, what caused him to turn to this particular tale—to elevate the narrative of a small ethnic group into the organizing principle of his model of human history, to use it to "construct a Pan-Asian consciousness"? and (2) besides providing a *model* of heroic behavior, emulated within the whole of "Otter's Children," is the Oroche myth reflected elsewhere in Khlebnikov's oeuvre on a thematic level: can one bring additional texts into the complex of works impacted by the myth?

An attempt to deal with the first issue was made by P. Tartakovskii (1987). For him, the key factor[3] is the role of the Oroche narrative as a model of the transformation of chaos into cosmos (an act emulated elsewhere in the "supersaga"): of bringing unity, symbolized by the single remaining sun, out of disorder. The striving towards world harmony has a moral dimension, while the many individuals who contribute to this effort gain a special status: "Anyone who brings goodness, harmony and justice to the world can become a Son of Otter" (1987: 206). Tartakovskii claims that this message of the Oroche myth is the Eastern component of Khlebnikov's attempted "Pan-Asian consciousness";[4] it would have held special significance for the poet as a product of a continuing mythological tradition, through which the present is linked with the mythic past (Tartakovskii 1987: 198–211).

This analysis, though argued on the basis of texts and various sources, must be questioned on two major points. First, it is doubtful that Khlebnikov saw the basic meaning of the Oroche myth in such putatively revolutionary terms; too many of the *dramatis personae* of "Otter's Children" are basically Romantic figures, whose striving is narrowly patriotic or nationalistic, rather than universalist. Second, the view of Khlebnikov's creation is excessively abstract: absent is the special quality of *poetic myth*, which implicates any poet's persona in his works and, on that basis, endows some of them with added emotional load and status. This crucial dimension of myth is certainly present in Khlebnikov.

My use of the term "poetic myth" echoes the ideas of Roman Jakobson, whose pithy formulation of the dialectic between myth and poetry is cited above.[5] The Jakobsonian conceptual framework is both flexible and sophisticated, and has been validated by the broad thrust of semiotic research. An important aspect of this methodology lies in looking at entire *episodes* rather than isolated elements regarded as mythological. By focusing on *action*, on *plots*, one can correlate the varied actants in similar situational units, and, indeed, compare and link seemingly disparate texts.[6]

Thus, the key to uncovering a fragment of a mythology lies in being able to trace a symbolically strong episode, a semantically marked situation through its varied implementations and transformations in a poet's oeuvre. As an initial step in the present discussion of Khlebnikov texts influenced by Oroche narratives we may consider the poem "Something Gloomy" ("Mrachnoe"). Where the initial sections of the "supersaga," as well as the closely-linked poem "The Fires" ("Plamena," 1912), conclude in a victory for the challenger of the sun, this short lyric suggests a different resolution:

When I get tired of myself,
I'll hurl myself into a golden sun,
I'll put on a rustling wing,
I'll mix together vice and what is holy.
I've died, I've died, and the blood has rushed
Down my armor in a wide flood.

I've woken up differently, again
Casting a warrior's glance at you.[7]

The poem presents a death and a resurrection—a vision consistent both with Khlebnikov's interest in metempsychosis and his theories of historical cycles. Yet though its message is ultimately heroic, with the reborn "I" "casting a warrior's glance at you" ("okinuv vas voina okom"), the brief narrative implicit in the poem's fictive world includes a *fiery death resulting from the hero's decision to act*. The situation in which the lyric "I" finds himself thus anticipates the following curious entry in his fragmentary diary:

14 iunia <1914> sozertsaia sebia v storone.

Novosti:

Khlebnikov iz neumolimogo prezreniia k sebe v 101 raz brosal sebia na koster i plakal, stoia v storone. (SP V: 328)

14 June <1914> contemplating myself from the sidelines.

News:

Khlebnikov, out of relentless contempt for himself, cast himself into the fire for the 101st time and wept, standing on the sidelines.[8]

The similarity of the two segments involves not only the death of the "I," presented in the poem with a shift in temporal perspective and in the diary entry as an object rather than subject, but also the nature of that death: by fire, a result of deliberate self-immolation. This suggests that in searching for other texts potentially connected to the Oroche myth—either prior or subsequent to the moment of Khlebnikov's encounter with Margaritov's volume—works be included where fire, an element from the same paradigmatic series as the sun, plays a central role.

Three earlier Khlebnikov texts fit this selectional criterion. One is the little-known play "Mystery of the Distant Ones" ("Tainstvo dal'nikh"), apparently created in 1908; another—an unpublished prose draft, undated yet also likely from this period; the third—a minor prose sketch from 1905, "There was darkness..." ("Byla t'ma...").

The time and place of "Mystery of the Distant Ones" are undetermined, though its cast of characters—including a priestess, a priest, youths, elders, warriors, etc.—suggests classical Greece. The brief work begins by depicting a crowd of young men obsessed with the priestess-queen who, as part of a rite, is ready to give herself to all who desire her. Her attraction is truly fatal: a number of youths, unable to wait for the proposed coupling, kill themselves. The proposed orgy is interrupted by nature, as a long-dormant volcano suddenly erupts. In response, the priestess changes the nature of the rite, proposing to embrace those only who will plunge an arm into the fiery lava. When one youth survives this test the priestess gives herself to him, urging the remainder to share in the erotic sacrament by imagining themselves in the place of the youth. In the end, the lava again flows, and all the participants in the "mystery" perish in its fires.

In the play, direct analogies are drawn between the world of man and the world of nature; throughout, imagery of the fire overwhelms the text. Initially, the sun, whose progress marks the progress of the day, is the object of entreaties from young men eager for the priestess' promise: "Sun! Sun! And we fall from high cliffs into the sea, praying to you: Sun! Sun! speed your course!" ("Solntse! Solntse! I s vysokikh skal padaem v more, umoliaia tebia: Solntse! Solntse! uskor' svoi beg!"). The flow from the volcano (a member of the same paradigmatic series as the sun) is compared to the "seed of a youth that knows no barriers" ("semia iunoshi, ne znaiushchee pregrad"). The priestess declares that "each one, who will reach the fiery monster and will ascend the hill, with his blackened hand as proof, will be taken by me into a sweet, fiery embrace" ("vsiakii, kto dobezhit do ognennogo chudovishcha i vzoidet na kholm, nesia dokazatel'stvom ispepelennuiu ruku, budet priniat mnoi v sladostnye, i ognennye ob"iatiia").[9]

How is this curious yet artistically weak text to be viewed? S. Kazakova suggests that it, as well as the better known "The Girl-God," was written under the influence of Nietzsche (via Viacheslav Ivanov) and that it requires a "Dionysian code" for its interpretation (1990: 440). Her argument is generally persuasive, and the parallel she adduces between solar imagery in "Mystery" and in another Nietzsche work, *Thus Spake Zarathustra,* is quite striking, and suggests that this text too influenced Khlebnikov in his writing the play.[10] Additionally, the close linkage of two themes, love and death, may also reflect the impact of Pushkin, especially his Egyptian Nights (Baran 1992: 367). The influence of Symbolism, especially of F. Sologub's novella *Queen of Kisses* (*Tsaritsa potseluev*) is also probable.

Yet besides these likely literary influences there remains the uncomfortable fact that "Mystery of the Distant Ones" is, at its root, deeply personal. This claim can be made without relying on psychological analysis; it is supported by our knowledge of the complex yet very palpable relationship between his biography and his art. In particular, an important factor in Khlebnikov's poetics was his tendency to project his own persona into his works, and, frequently, to appear in them in the guise of various doubles.[11] We are justified in suspecting that this tendency is operative in the present instance, and that the play represents a clearcut—and a very puerile—example of the poet's transformation of his erotic fantasies and fears into literary form. *Since its plot—whereby the characters, driven by Eros, perish in a confrontation with a fiery enemy—is quite similar to the basic pattern found in the Oroche tale, it is likely that it was an important precursor to the borrowing and reconstitution of the myth.*

Sexual drive is not the only generator of a plot situation analogous to the Oroche mythic text. Another was political concerns, which in the poet's case prior to World War I meant nationalistic, Slavophile, and even, to some extent, Black Hundred-type views. A reflection of these is found in the following incomplete prose draft (presented below in both the original and in translation) which has been preserved, along with a number of other fragments of his prose, in the small Khlebnikov archive in the Saltykov-Shchedrin Library in Leningrad:

Slusha<ite> slush<aite> tam sobralis' bare—odety v nerusskie <odezhdy>—prodavat' Rossiiu.

Oni ee deliat na chasti kak

My stuchalis' nas ne pustili. My khoteli kriknut' nas ne poslusha<li>.

Idem podzhech' ikh.

Dukh s sel: I ia s vami. Podzhech'! Podzhech'!

Begut. Iz pereulka vybegaet neskol'ko liudei v otrep'iakh i poliv sebia chernym maslom kidaiut<sia> kak zazh<zhennye> sveto<chi> v dveri. Gustoi dym, plamia. Kriki.

Vybegaiut gospoda s zamazannymi edoi gubami, ili s pal'tsami belymi ot mela.

Vyskakiv<aet> molodoi chelovek i.... roniaet vstavnuiu cheliust'.

Smeshlivyi molodoi chelovek: Poslusha<ite>, ne pravda-li eta cheliust' napominaet bratstvo, ravenstvo i svobodu.

Dukh: O Um, o Um! Nishchie i vory spasli zemliu. Oni vozstali, oni pobedili.

Bogovaten': Da budet blago<slovenno> va<she> plemia vo veki vekov![12] Listen, listen, there the gentry have gathered—dressed in non-Russian clothes—in order to sell Russia.

They divide her into part like ...

We knocked but weren't let in. We wanted to shout out, but were not heard. Let us go and set fire to them.

Spirit from the villages: I too am with you. Let's set fire! Let's set fire!

They run. From the sidestreet run out several people in rags and, after pouring black oil on themselves, throw themselves through the doors like lit torches. Thick smoke, flames. Shouts.

Gentry run out: they have remnants of food on their lips, or else their fingers are white from chalk.

A young man runs out and.... drops his false teeth.

A funny? young man: Listen, isn't it true that these false teeth resemble fraternity, equality, and liberty?

Spirit: O Mind, o Mind! Beggars and thieves have saved the land. They rose up, they triumphed.

Divinity: May your tribe be blessed forever and ever.

Although the draft is undated, the inclusion of a divinity (*bogovaten'*) and a "spirit from the villages" (*dukh s sel*), as well as its subject matter—an imagined suicidal attack on those seen as betraying Russia, echo the cast of characters and thematics of Khlebnikov's 1908 coinage-rich play "Snowhite" ("Snezhimochka"), suggesting that the prose text was likely created at about the same time. The targets of the self-immolation are members of the Westernized elite, presented with the type of grotesque detail (false teeth) that is found elsewhere in Khlebnikov's pre-World War I writings, e.g. the story "Easter Day" ("Velik-den'") and the play "The Marquise des S." ("Markiza Dezes"), though the degree of bitterness found in the draft is far more explicit than in the published works.[13] In addition, the motif of surrender to foreign interests while manipulating radical political slogans links the "gentry" in this fragment with another group, the Jews, anathemized by the poet in some other draft texts.[14]

The denouement of this brief scene bears a striking resemblance to the core of the cosmogonic myth: although a collective protagonist is involved, and the sun is replaced by a fire, the essence of the situation—a willingness to confront, and be consumed by, the flames—remains the same.[15] Khlebnikov's creation of a fantastic scene of self-immolation is all the more striking because, setting aside the practices of the Old Believers, there was a lack of historical precedents for such an action.[16] Once again, the poet has constructed an episode of fiery death to express a deeply held idea, and he would certainly have recalled his own draft when encountering the Oroche story of mankind's first days.

Self-immolation is also the high point of a prose miniature, the earliest of the three works. The allegorical "There was darkness...," which owes much to Gor'kii's "Song of the Falcon" ("Pesnia o sokole"), depicts a land covered with darkness and peopled by beings whose primary feeling is one of boredom. A change is brought about through Romantic action:

And in this darkness there was a firefly who had this thought: 'What is better: to crawl about for a long, long time in darkness and a life that is unnoticeable, or to light up once with a white fire, to fly as a white spark, and as a white song tell of another life, not of black darkness but of a game and of streams of white light.' And he thought no further but covered his delicate wings with pitch and willow fuzz, and, enflamed and chased by raging fire, pitiful and small, flew as a white spark in the black darkness and fell with singed wings and legs, frightened, dying.

And the black darkness crushed with specters the firefly—in his final troubled moments laying there delirious, his imagination in a fever.

But the light gleamed. And sight came to the beings in darkness, who quietly and stickily crawled along the ground: longing for light woke with a swan's power. (Utes: 41–42)

Though the code used here is basically ornithological, the mythological core of the work is essentially the same as in the other texts already examined. On the one hand, the protagonist of this episode engages in a fiery self-sacrifice; on the other, he may be viewed as a life-affirming culture hero, albeit on a microcosmic, rather than cosmic scale. However one looks at this text, its status as a predecessor to the Oroche solar myth is beyond doubt.

The early texts create a thematic background into which the Oroche myth of the culture hero challenging the three suns could be easily absorbed and reinterpreted. Significantly, the theme of heroic immolation is reiterated within "Otter's Children" itself, in the figure of Jan Hus, one of the participants in "dialogue of the dead" in Part VI of the "supersaga." (CW II: 306)

Analogues to the mythic situation are found in two additional works, written between 1911 and 1913, that is, during the same period as "Otter's Children." First, the theme of a fiery self-sacrifice allows us to bring into the purview of the heliomachic myth the charming primitivistic poem "I and E" ("I i E"). Subtitled "Tale of the Stone Age," it involves two protagonists, a young woman and young man, who travel away from their home tribe and fall into the hands of a cruel neighboring tribe, whose priests sentence them to be burned alive for blasphemy. The young couple is tied to the stake, the fire is set, but the two are saved through divine intervention, and are asked by their kinfolk to return as rulers. The poem emphasizes the idea of the couple's bold deed: "We fulfilled a debt of boldness,/ A deed proud and stern" ("My svershili smelyi dolg,/ Podvig gordyi i surovyi"—Tvoreniia: 203). Significantly, Khlebnikov's unusual Postscript to the poem, which summarizes and interprets its plot, emphasizes the meaning of the sacrifice and miraculous salvation: "In this way, through deed, through the fire lay their path to rule over their kinfolk" ("Takim obrazom, cherez podvig, cherez ogon', lezhal ikh put' k vlasti nad rodnymi"—Tvoreniia: 204).

The second work that bears on our concern is the impressionistic prose sketch "Lubny is a strange, godforsaken town" ("Lubny—svoeobraznyi glukhoi gorod..."). This brief work describes a small provincial town in Poltava province where Khlebnikov's family lived for a time. Two connected key motifs in the narrator's rendition of Lubny are those of the frequent fires and of the sound of the fire brigade's bugle: combined, these evolve into an apocalyptic vision of the Last Judgment and of the flames which ultimately await mankind as a whole:

> ... These bugles ignore you and your private passions; they know only the people as a mass and they twist its will like a serpent, as they hurl themselves forward to conquer fire.
>
> 'Wake up,' they cry, 'fire is loose, go put it out, bind and chain it and throw it back in its cage. Its time has not yet come, *the final struggle between man and fire has yet to be joined* (my italics—H.B.). It is not yet time to tame the beast.'
>
> I thought for a long time about the immeasurability of their grandeur. I knew that all things that exist are only written signs and I have made constant efforts to understand them, for, after all, a grasp of number is the great translator between languages that bear no relationship to one another. In these sounds, agonizing and threatening, speaking a language of some kind or other, one could feel the breeding place of the resurrection of the dead. And in the terrifying howl that rose at an angle above the world, and fell back onto it like lava from heaven, was hidden the promise of a day when fire would be victorious, a precursor and sign dear to the hearts of the people. *Is fire the natural state of the deceased? Are the deadly embraces of the sun so distant?* (my italics—H.B.).[17] For things living are more akin to this earth than things dead. And the combat of fire and earth with fire victorious, ripping open the covers of earthly graves and burning them up, that is what [illegible] disturbs you after [illegible].
>
> And some day he will come, this angry crimson conqueror, this red fire.
>
> If, in death, our mortal wax bids farewell to fire, then what we hear at such moments is the return of mankind as a thing of fire.... (CW II: 37)

Here, the myth comes full circle—the culture hero's fiery contest in the first days, perhaps alluded to in the phrase "the sun's death," is replaced by the imagined final challenge, this time to mankind as a whole. This projection into the future, envisioning "the return of mankind as a thing of fire," shows both the potential for evolution of the mythic complex and its appeal for Khlebnikov.

The several analogues to the Oroche solar myth all date from the time before the First World War. This is the period when, as has been suggested previously, Khlebnikov's works gravitate towards antithetical constructions, frequently manifested on the level of both plot and theme (Baran 1985). Subsequently, the poet's own experiences as a recruit and what he witnessed on his many wanderings during the Revolution and the Civil War lead him to deromanticize conflict and lessen the emphasis on heroic personalities such as Son of Otter: Khlebnikov's mythopoesis now underscores resolution of conflicts, neutralization of oppositions. Not surprisingly, a rare example from this period of what might be regarded as a variation on the basic myth has very different ideological underpinnings and a different result. It is found in the poem "The One, the Only Book" ("Edinaia kniga") which in 1920 became the beginning of the "supersaga" "Azia Unbound" ("Azy iz uzy"):

> I have seen the black Vedas,
> the Koran and the Gospels
> and the books of the Mongols
> on their silken boards—
> all made of dust, of earth's ashes,
> of the sweet-smelling dung
> that Kalmyk women use for morning fuel—
> I have seen them go up to the fire,
> lie down in a heap and vanish
> white as widows in clouds of smoke
> in order to hasten the coming
> of the One, the Only Book,
> whose pages are enormous oceans
> flickering like the winds of a blue butterfly,
> and the silk thread marking the place
> where the reader rests his gaze
> in all the great rivers in a dark blue flood: (CW II: 322)

Like the phoenix, Khlebnikov's Book arises from the ashes of the old religions. Similarly, a new myth—of reconciliation, of universal harmony, of the unity of Culture and Nature—replaces the old mythic plot. In method as well as contents, the passage strikingly confirms Boas' oft-cited words, "it would seem that mythological worlds have been built up, only to be shattered again, and that new worlds were built up from the fragments."[18]

Both of the questions posed earlier have been answered in the affirmative. A key reason for Khlebnikov's affinity for the Oroche story of the culture-hero challenging two of three suns was its similarity to a narrative situation he had conceived previously and had used in texts with deep personal overtones for him. The myth—an external, collective creation—served to validate private obsessions, and fused with them into a central component of Khlebnikov's poetic mythology.

Notes

1. I have followed the text of the English translation of *Besedy* with one emendation: "stikhii" has been translated as "elemental," rather than "elementary forces." For the Russian original of this passage—both more precise and more forceful—see Jakobson (1988: 557).
2. My identification of this source has generally been accepted (e.g., Tvoreniia: 691; Ivanov 1986: 427–428; Tartakovskii 1987: 198), though some curious assertions about Khlebni-

kov's possible readings in this area persist in the scholarly literature. Thus, P. V. Popov asserts that Margaritov's volume "does not include a cosmogonic myth," and suggests, after examining various possibilities, a hypothetical encounter between Khlebnikov and the anthropologist L. Ia. Shternberg or one of his colleagues in the apartment ("Tower") of the Symbolist maitre Viacheslav Ivanov (Popov 1987: 131–132).

3. Tartakovskii also notes the possibility of Khlebnikov's having learned a number of Siberian mythological narratives from his father, who had some firsthand knowledge of the native cultures; the poet's interest may also have been sparked by the fact that in Kazan', where he studied for a while, a number of works on the Oroches had been published (Tartakovskii 1987: 198–200).
4. An equivalent Western myth revolves around the figure of Prometheus, introduced in the "supersaga" in Part V.
5. Jakobson's conception was laid out in detail in his seminal paper "The Statue in Pushkin's Poetic Mythology" (1937), where he not only analyzed in detail a fragment of Pushkin's symbolic system, but sketched out certain methodological concomitants for conducting this type of study. Discovery of a poetic mythology centers on constants, on *semantic invariants*; these function as a *system*, be it within a cycle or the poet's entire oeuvre, and must be studied in their relationship to other elements of the poet's work; they must also be considered in relation to potential *biographical data*, to the "situation" which may have prompted the creation and specific shaping of a text. For Jakobson, the biographical event is an active element of the work of art and hence must be brought into a discussion of poetic myth: "The situation is a component of speech; the poetic function transforms it like every other component of speech, sometimes emphasizing it as an efficient formal device, sometimes, on the contrary, subduing it, but whether a work includes the situation positively or negatively, the work is never indifferent to it" (Jakobson 1987: 320). The manner in which biographical events interweave with elements of literary works to form a poet's mythological system is illustrated further by Jakobson in a series of studies, ranging from discussions of Pasternak and Maiakovskii to the Czech Romantics Mácha and Erben. (Several of these papers are included in Jakobson [1987].) Throughout, Jakobson emphasizes also the *potential for variability* among individual instances of implementing the mythic pattern, which sometimes coincide and at other times are in a relationship of complementary distribution, yet their totality confirms the invariant that bears a symbolic load within a particular author's life and writings.
6. Such an approach avoids, for example, the problems found in a recent discussion of Khlebnikov's use of solar symbolism in various works—esp. "Otter's Children" —where the author emphasizes traditional associations of the image of the sun in different cultures and uses them to analyze the ideological level of the works in question (Garbuz 1986). Unfortunately, the investigator focuses too greatly on the extratextual background of individual symbols without looking at their role on the level of plot. In one case, he draws on the opposition between the light and the dark principles in Iranian mythology (Ahura Mazda vs. Angra Mainyu) to interpret the mythic narrative found at the beginning of "Otter's Children," and, in line with this dichotomy, "locates" in the "supersaga" a protagonist who does not exist. Furthermore, Gorbuz sees the sun as providing a basis for treating three Khlebnikov texts—"Otter's Children," the play "The Girl-God" ("Devii bog"), and the prose tale "Ka"—as a single work, a "Pan-Asian" "supersaga" on a larger scale. Although the poet, late in life, considered creating such a work, it is doubtful whether solar thematics would have been paramount in his mind: the role of the sun in "The Girl-God" is largely decorative, while in "Ka," where it is linked to the figure of the Egyptian religious reformer Pharaoh Akhnaton (Amenhotep IV), it still plays a less decisive function than in "Otter's Children."
7. Here and elsewhere, unless otherwise indicated, translation is mine—H. B.
8. This connection was first noted by J. Döring-Smirnov and I. Smirnov (1986: 15–16).

9. Quotations taken from S. Kazakova's recent publication (with some errors!) of the play (Khlebnikov 1990), which is based on the autograph in kept in TsGALI, *fond* 527, *op.* 1, ed. *khr.* 100.
10. This is in spite of a basic mistake in her reading of the text, where she treats lines referring to the volcano as referring to the sun. On Khlebnikov's response to Nietzsche, see also Baran (1993).
11. See, for example, the extensive study by R. Vroon (1986).
12. OR GPB, *fond* 1087, *ed. khr.* 17, *fragment* 5. For a description of the Khlebnikov holdings, along with the texts of a number of previous unknown early prose works, see Zubkova (1988).
13. 'In "Easter Day," for example, which is set during the holiday celebration in a Ukrainian village, there is this ironic characterization of the local intelligentsia: "Meanwhile the local Social Democrats, men and women, settled like sparrows on the house benches and twittered about Kautsky, just like sparrows in good weather. As she passed by them, the girl darted angry glances. 'Good-for-nothings,' she muttered." (CW II: 32).
14. Compare this transcription of a portion of another draft fragment (variants omitted), where Khlebnikov voices his outrage that the issue of anti-Jewish pogroms—raised by foreign powers in connection with policy towards Russia—may affect his country: "Drevnii mudrets iskal meru vsekh veshchei. / Ia nashel emu iz zhelchi i pravdy otvet: evrei. / Potomu-chto sud'by ne samoi-li Rossii resha<iu>tsia v savisi<mosti> ot-togo / byl-<li> pushchen povsiudu pukh iz perin<y> griaznoi evreiki...." ("An ancient sage sought the measure of all things / I have found for him an answer—the Jew—out of bile and truth. / For is not the fate of Russia itself decided on the basis of whether / Feathers flew out of the quilt of a dirty Jewess....") (OR GPB, *fond* 1087, *ed. khr.* 17, *fragment* 12). Compare the anti-Jewish passage in the "1908 Notebook" (Baran 1991: 96-97). Among the published works, there is the play "Snowhite," which contains a few positive mentions of the Black Hundred movement that were excised its original publication (NP: 64–75).
15. An echo of this plot, though transformed into a less violent situation, is found in the words of one of the participants in the dialogue "Two Individuals" ("Razgovor dvukh osob"): "I sometimes dream of a great bonfire of books. Yellow sparks, white flames, and transparent ash—an ash so fine that the touch of a finger, a breath even, will cause it to crumble—though even then a few arrogant, boasting lines can still be read upon it. Then the whole mass sinks down and becomes a beautiful black flower, its heart still aglow from the fire, grown from a book written by human beings, just as the fires of nature grow from the book of Earth—or from lizards' crests and other such fossils." (CW I: 288). Here, as in the comic poem "Malusha's Granddaughter" ("Vnuchka Malushi"), the proposed incendiary act would affect not people but the books which ultimately shaped what are seen as their pernicious ideas.
16. The sacrifices of Buddhist monks in South Vietnam and of Jan Pálach lay in the distant future.
17. "Ognevaia li priroda usopshikh, dal'nie li ob"iatiia smert<i> solntsa?" (Tvoreniia: 514). Compare my own translation, which I believe is closer to the Russian original: "Is it the fiery nature of the dead, the distant embraces of the sun's death?"
18. Quoted in Lévi-Strauss (1973: 21).

Works Cited

Primary Sources:

Khlebnikov, V. 1928–33. Sobranie proizvedenii. 5 vols. Ed. N. L. Stepanov. Leningrad: Izdatel'stvo pisatelei [abbreviated—SP, with volume number].

___. 1940. *Neizdannye proizvedeniia*. Ed. N. Khardzhiev and T. Grits. Moscow: Khudozhestvennaia literatura [abbreviated—NP].
___. 1986. *Tvoreniia*. Ed. V. P. Grigor'ev and A. E. Parnis. Moscow: Sovetskii pisatel' [abbreviated—Tvoreniia].
___. 1987. *Collected Works of Velimir Khlebnikov*. Vol. 1. *Letters and Theoretical Writings*. Trans. Paul Schmidt. Ed. Charlotte Douglas. Cambridge, Mass. and London: Harvard University Press [abbreviated—CW I].
___. 1988. *Utes iz budushchego. Proza, stat'i*. Ed. R. V. Duganov. Elista: Kalmytskoe knizhnoe izdatel'stvo [abbreviated—Utes].
___. 1989. *Collected Works of Velimir Khlebnikov*. Vol. 2. *Prose, Plays, and Supersagas*. Trans. Paul Schmidt. Ed. Ronald Vroon. Cambridge, Mass. and London: Harvard University Press [abbreviated—CW II].
___. 1990. "Tainstvo dal'nikh." *Russian Literature* 27: 453–458.

Secondary Sources:

Baran, H. 1973. "Xlebnikov and the Mythology of the Oroches." In: *Slavic Poetics: Essays in Honor of Kiril Taranovsky*. Ed. R. Jakobson et al., 33–39. The Hague–Paris: Mouton Publishers.
___. 1976. *Xlebnikov's "Deti Vydry": Texts, Commentaries, Interpretation*. Ph.D. diss., unpubl. Harvard University.
___. 1983. "Temporal Myths in Xlebnikov: From 'Deti Vydry' to 'Zangezi'." In: *Myth in Literature*. New York University Slavic Papers 5. Eds. A. Kodjak et al., 63–88. Columbus, Ohio: Slavica Press.
___. 1985. "Xlebnikov's Poetic Logic and Poetic Illogic." In: *Velimir Chlebnikov. A Stockholm Symposium*. Acta Universitatis Stockholmiensis // Stockholm Studies in Russian Literature 20. Ed. N. A. Nilsson, 7–25. Stockholm: Almqvist & Wiksell International.
___. 1991. "V tvorcheskoi laboratorii Khlebnikova: O 'Tetradi 1908 g.'." In: *Zaumnyi futurizm i dadaizm v russkoi kul'ture*. Ed. Luigi Magarotto et al., 89–101. Bern: Peter Lang.
___. 1992. "Pushkin in Khlebnikov: Some Thematic Links." In: *Cultural Mythologies of Russian Modernism: From the Golden Age to the Silver Age*. California Slavic Studies 15. Eds. B. Gasparov et al., 356–381. Berkeley: University of California Press.
___. 1993. "Khlebnikov and Nietzsche: Pieces of an Incomplete Mosaic." To appear in volume of studies from 2nd "Nietzsche in Russia" conference. Ed. Bernice Rosenthal [in press].
Döring-Smirnov, J. R. and I. P. Smirnov. 1986. "Der Futurismus Chlebnikovs." In: *Velimir Chlebnikov 1885–1985*. Sagners Slavistische Sammlung Band 11. Eds. J. Holthusen et al., 9–29. München: Verlag Otto Sagner.
Garbuz, A. V. 1986. "Solnechnaia simvolika v mifotvorchestve V. Khlebnikova." In: *Fol'klor narodov RSFSR. Epicheskie zhanry, ikh mezhetnicheskie sviazi i natsional'noe svoeobrazie. Mezhvuzovskii nauchnyi sbornik*. Eds. T. M. Akimova and L. G. Barag, 121–129. Ufa: Bashkirskii gosudarstvennyi universitet.
Ivanov, Viach. Vs. 1986. "Khlebnikov i nauka." In: *Puti v neznaemoe: Pisateli rasskazyvaiut o nauke. Sbornik 20*, 382–440. Moscow: Sovetskii pisatel'.
Jakobson, R. 1937. "The Statue in Puškin's Poetic Mythology." In: Jakobson 1987: 318–367.
___. 1987. *Language in Literature*. Eds. Krystyna Pomorska and Stephen Rudy. Cambridge, Mass. and London: Harvard University Press.
___. 1988. *Selected Writings*. VIII. *Completion Volume One*. Ed. Stephen Rudy. Berlin–New York–Amsterdam: Mouton de Gruyter.
Jakobson, R. and K. Pomorska. 1983. *Dialogues*. Trans. Christian Hubert. Cambridge, Mass.: MIT Press.
Kazakova, S. Ia. 1990. "*Tainstvo dal'nikh*—'Dionisicheskaia' p'esa Velimira Khlebnikova." *Russian Literature* 27: 437–451.
Lévi-Strauss, C. 1973. *The Savage Mind*. Chicago: The University of Chicago Press.

Margaritov, V. P. 1888. *Ob orochakh imperatorskoi gavani*. St. Petersburg.

Popov, P. V. 1987. "V. Khlebnikov—myslitel' i poet. (Vremia i prostranstvo kak mifologemy ego tvorchestva)." In: *Vliianie nauki i filosofii na literaturu*. Ed. N. V. Zababurova, 124–137. Rostov-na-Donu: Izd. Rostovskogo universiteta.

Tartakovskii, P. I. 1987. *Sotsial'no-esteticheskii opyt narodov Vostoka i poeziia V. Khlebnikova. 1900–1910-e gody*. Tashkent: Izd. "FAN" Uzbekskoi SSR.

Vroon, R. 1986. "Metabiosis, Mirror Images and Negative Integers: Velimir Chlebnikov and his Doubles." In: *Velimir Chlebnikov (1885–1922): Myth and Reality*. Studies in Slavic Literature and Poetics VIII. Ed. Willem G. Weststeijn, 243–290. Amsterdam: Rodopi.

Zubkova, N. A. 1988 "Iz rannei prozy V. V. Khlebnikova (Po materialam fonda Otdela rukopisei Publichnoi biblioteki)." In: *Issledovaniia pamiatnikov pis'mennoi kul'tury v sobraniiakh i arkhivakh Otdela rukopisei i redkikh knig: Sbornik nauchnykh trudov*. Ed. L. I. Buchina et al., 151–178. Leningrad: GPB im. M. E. Saltykova-Shchedrina.

SLAVIC ARCHIVES: EAST AND WEST

Tiutchev's Political Memorandum Rediscovered (Preliminary Notes)

ALEXANDER OSPOVAT

University of California, Los Angeles

AFTER F. I. TIUTCHEV'S forced resignation from the post of secretary-general of the Russian diplomatic mission in Turin in 1839 (Lane 1987), he remained in Europe, living a private life in Munich, without ever losing hope of renewing his career in the diplomatic corps. In his efforts to seek support for reinstatement from various high-ranking persons, he relied especially on the mediatory efforts of an old friend, Amalia Krüdener. From the mid-1830's this Saxon beauty, who was first the mistress of Nicholas I and later of Count A. Kh. Benkendorf, the chief of the Secret Police (which was known as the Third Department), had enjoyed considerable influence in St. Petersburg. In 1839, she was described in passing by the American Ambassador to Russia as "Madame Krüdener, decidedly recognized beauty and a great favorite" (*Diary of G. M. Dallas* 1970: 177). We believe that it was due to Amalia Krüdener's patronage that Tiutchev received his initial appointment in Turin in the summer of 1837 (Ospovat 1985: 71–72). Significantly, her absence from Russia in the autumn of 1842 induced Tiutchev to postpone his trip home for a year (Tiutchev 1984, 2: 69).

In June of 1843, Tiutchev arrived in Moscow from Munich, and came to the capital on approximately August 24 (September 5). Writing to his wife around August 28 (September 9), he informed her that Benkendorf had invited him and the Krüdeners to his Falle estate, located near Revel (Tiutchev 1984, 2: 88). At the end of that five-day visit, Tiutchev sent a letter to his parents, in which he noted the interest shown by the Chief of the Third Department in the "proposal known to you," and his "readiness to present these ideas to the Emperor" (Aksakov 1886: 30). Somewhat later, in a letter to his wife written around September 15 (27) or September 17 (29), Tiutchev re-

Elementa, 1993
Vol. 1, pp. 91–97
Reprints available directly from the publisher
Photocopying permitted by licence only

marked: "But I am not as grateful to him [Benkendorf] for his hospitality, as I am for conveying my ideas to the Emperor, who paid more attention to them than I had ever dared hope" (*Starina i novizna* 1914, 18: 11).

Until now, Tiutchev's proposal had not been located, and researchers had attempted, at most, to reconstruct only the general sense of that political initiative (Aksakov 1886: 29; Pigarev 1962: 111). The most detailed, and, as it appears to us, the most plausible of these reconstructions belongs to E. P. Kazanovich (1928: 162–163), who wrote:

> Expounding his views on Russia's providential role in what once used to be known as the Byzantine East ... Tiutchev probably presented a detailed account of the Western position on the Eastern question to the Emperor, pointing out the disadvantages of such views for Russia...and, in conclusion,... proposed that efforts be made to peacefully influence Western diplomatic policies and public opinion, by persuading powerful representatives of the Western press and specialists to take the Russian side."

Agreement was reached at the Falle meeting that a final decision on Tiutchev's proposal would be made a year later (Aksakov 1886: 30). However, Benkendorf died in September 1844, and only in 1845, after his return to Russia, did Tiutchev again attempt to draw the Emperor's attention to his political ideas. On September 17, 1845 Aleksandr Turgenev wrote his brother Nikolai, then in Paris, that he was sending him "Tiutchev's famous memorandum to the Emperor.... He received 6,000 rubles *Wartgeld* and now is being promised a diplomatic posting. The blotted out lines at the end of the memorandum contained a fairly clear hint at how useful he might be, were he to be entrusted with editing articles on Russia for foreign newspapers."[1] This copy of the text was finally located in the St. Petersburg archives of the Turgenev brothers.[2] The archive document is a clerk's copy, which lacks both a title and the author's name, but includes an accompanying note in Aleksandr Turgenev's own handwriting (as this note cannot be comprehended outside of its context, it will be quoted later).

In order to dispel all doubts about the authenticity of the Turgenev text, we thought it expedient to continue our search in the Moscow archives of M. P. Pogodin, since in his diary one finds the following entry for June 15, 1845: "In the morning, Tiutchev [arrived]. [We spoke] about politics. He brought me his memo" (Kuzina 1989: 14; Azadovskii and Ospovat 1989: 69). This search was carried out by our colleague K. Iu. Rogov, who had located two more copies of the Tiutchev memorandum in Pogodin's archive.[3] The "Pogodin" copies, which are not differentiated in any way, are also completely identical to the "Turgenev" copy of the text.

The memorandum, written in French in the year 1845, is one of Tiutchev's earliest political compositions. A. Turgenev has noted the correlation of this text with Tiutchev's brochure entitled, "A Letter to Dr. Gustave Colb, Editor of the Universal Gazette"[4] (Munich 1844 [Tiutchev 1913: 333–343]). A retrospective look at the text, however, allows one to recognize it as a very succinct but most clearly formulated sketch of Tiutchev's personal philosophy of history, which is expounded in its fullest in his unfinished treatise "Russia and the West" (Tiutchev 1988).

Tiutchev's proposal takes up about seven pages of standard printed page format. It will be published elsewhere in full (both in the language of the original and in Russian translation), and with all necessary comments, but here we offer several key excerpts from that text:

"... What is really going on between the West and ourselves? Is the West sincere in its misunderstanding of us? Can it seriously claim ignorance of our historical rights?

Long before Western Europe was settled, we existed, and existed gloriously. The difference is that back then we were known as the Eastern Empire, or as the Eastern Church. What we were then, however, we remain today.

What is the Eastern Empire? It is the legitimate and direct descendent of the supreme authority of the Caesars. It is a fully and entirely sovereign state; in contrast to Western monarchies, it neither derives nor emanates from any outside power, but contains the justification for its authority within itself, while being regulated, contained, and sanctified by Christianity.

What is the Eastern Church? It is the Universal Church.

These are the only two issues on which any serious polemics between the West and ourselves can be based. All the rest is idle talk. The clearer our understanding of these two issues, the stronger our stand before our adversaries will be and the truer we shall become to ourselves. All things considered, the struggle between the West and ourselves never ended. Nor has there ever been a truce, but, at best, only brief cease-fires. It will do no good now to pretend otherwise: this struggle will soon break out with greater force than ever, and then, as it was previously, as it has always been, the Roman Church--the Latin Church, will head the enemy ranks.

So then, let us enter the battle openly and resolutely. When facing Rome, let not the Eastern Church forget for a single moment that she is the legitimate descendent of the Universal Church.

... Catholicism was always the force behind Papism, just as Papism was always the weakness of Catholicism. Only the Universal Church possesses strength without weakness. Let it appear, let it intervene in the debate, and we shall witness in our time what was apparent in the first days of the Reformation, when the leaders of that religious movement severed all ties with the Pope, but still hesitated to cut themselves off from the traditions of the Catholic Church; when they unanimously turned to the Eastern Church. Today, as in those days, religious reconciliation will come only from the Eastern Church; she holds the future of Christianity in her hands.

Such is the first, the loftiest issue which lies at the source of our dispute with Western Europe and it is a question of life or death. But there is yet another serious issue which is commonly called the Eastern question. And that is the question of the Empire.

... What does History tell us? It tells us that the Orthodox East—that immense world which took up the Greek cross—embraces common principles; that all its parts are united in solidarity, that it lives its own distinctive and indestructible life. It can be rent apart physically, but spiritually it will always be united and indivisible. For a short while it was subjected to

Latin domination; for centuries—to the invasion of the Asiatic tribes, but it never yielded to either force.

... Who can fail to see that the West, with all its philanthropy, all its pretended respect for the rights of nations, and all its wrath at Russia's insatiable ambitions, views all the peoples of the Turkish Empire merely as easy prey.

... Fortunately, historical Providence, which governs all earthly affairs, has not abandoned us. As mutilated and weakened as the Eastern Empire was in the thirteenth century, it had already managed to find enough strength to cast off the Latin domination after more than sixty years of troubled times. And, certainly, it must be acknowledged that since that time, the true Eastern Empire, the Orthodox Empire, that once fell has gloriously risen up.

... The great question ... will be settled here, amongst ourselves, in the very center, at the very heart of that world of Eastern Christianity, that Eastern Europe which we represent, that world which we comprise. Its ultimate destiny, which is also our own destiny, depends on ourselves alone. It requires, first of all, that we resolutely acknowledge the ties which bind us and set us apart.

... Once again let us repeat that if the West is hostile toward us, if they view us with ill will, it is because of an absurd, yet widespread opinion: while recognizing, perhaps even overestimating, our material power, they doubt that our might is embued with its own moral consciousness, with its own historical consciousness. Man is so created, especially the man of our time, that he will submit to physical force only if it is based upon the force of morality.

... Ours is an Empire, which by design or chance, knows no historical equal; it represents two immense entities: the fate of an entire race, and the best and the most sacred half of the Christian Church.

And yet, one still finds people who seriously question whether this Empire has the right to what it claims, whether it occupies a legitimate place in the world!.. Has this generation wandered so far into the shade of the mountain that it can no longer perceive its summit?... Is it at all surprising that the bearers of old convictions strive with all their might to suppress evidence which shatters and *supersedes* them? Should we not bolster this evidence, make it invincible and inescapable?[5]

At this point Tiutchev turns to the pragmatic aspect of his proposal. In his opinion, it is necessary to found a European press agency or to suborn an already existing agency, whose members are authoritative representatives of Western public opinion, around which a pro-Russian lobby could work. The *sine qua non* for directing "these separate, and up to a point, independent forces" is, in the author's opinion, a skilled organizer: "an intelligent man with strong nationalistic sentiments, profoundly devoted to the Emperor, and experienced enough in matters of the press to have acquired a thorough knowledge of the territory on which he will have to act." This role the author reserves for himself, writing: "Should this idea prove to be agreeable, I would be only too happy to lay before the Emperor all that a man can possibly offer and promise: purity of intentions and the zeal of absolute devotion."[6]

In the "Turgenev" copy, this last paragraph of the text is blotted out with ink, and there is a note in Russian beneath, in A. Turgenev's handwriting: "In the lines blotted out, T. hints that he should be chosen for the editorship of the articles concerning Russia." This note, when compared with the previously cited letter of September 17, 1845 from Aleksandr Turgenev to Nikolai Turgenev, provides us with ample evidence to recognize the text located in the Turgenev and the Pogodin archives as a copy of Tiutchev's proposal, submitted by him to the Emperor Nicholas I in 1843.

Even though this text met with the Emperor's approval, as is evident from Tiutchev's letter to his wife quoted earlier, no attempts were made to implement the author's proposal. Of course, the idea of co-opting a European newspaper must have impressed both Nicholas I and Benkendorf, but the scheme was much too bold, and provided no insurance, for instance, against possible diplomatic complications. Tiutchev was reinstated in the diplomatic service, but neither then nor later did Emperor Nicholas I's pragmatic regime call on his ideas and suggestions.[7]

Notes

1.. Archive of the Institute of Russian Literature (Rukopisnyi otdel Instituta russkoi literatury); hereinafter referred to as RO IRLI: *fond* 309, #950, reverse of page 334. On September 9 (21) of 1845, Aleksandr Turgenev, who had only recently returned to Russia, wrote to his brother: "Together with [N. V.] Sushkov, I had paid a visit to his wife, née Tiutchev [Daria Ivanovna, the poet's sister]: there I was given the French article written by her brother ... I have duly read it, and am now seething with indignation...." (RO IRLI: *fond* 309, #950, reverse of p. 333). And on September 22 (October 4) of 1845, Aleksandr Turgenev recorded the following entry in his diary: "I have sent off ... Tiutchev's article to my brother" (*ibid.*: #300: 46).
2. RO IRLi: *fond* 309, #2301.
3. Archive of the Lenin State Library (Rukopisnyi otdel Gosudarstvennoi biblioteki im. V. I. Lenina): *fond* 231/III, 17.41.
4. See his letter to V. A. Zhukovskii from November 13 (25), 1845 (RO IRLI: *fond* 309, #28279: 16).
5. The original Tiutchev memorandum reads as follows:

"En effet de quoi s'agit il entre l'Occident et nous? Est-ce de bonne foi que l'Occident a l'air de se prendre sure ce que nous sommes? Est-ce sérieusement qu'il prétend ignorer nos titres historiques?—

Avant que l'Europe occidentale ne se fût constituée, nous existions déjà et certes nous existions glorieusement. Toute la différence c'est qu'alors on nous appelait l'Empire d'Orient, l'église d'Orient; ce que nous étions alors, nous le sommes encore.

Qu'est-ce que l'Empire d'Orient? C'est la tranmission légitime et directe du pouvoir supréme, de pouvoir des Césars. C'est la souveraineté pleine et entière, ne relevant pas, n'èmanant pas, comme les pouvoirs de l'occident, d'une autorité extérieure quelle qu'elle poisse être, portant son principe d'autorité en elle-même, mais reglée, contenue et sanctifiée par le Christianisme.

Qu'est-ce que l'église d'Orient? C'est l'église universelle.

Voilà les deux seules questions sur lesquelles doit rouler toute polémique sérieuse entre l'Occident et nous. Tout le reste n'est que du verbiage. Plus nous nous serons pénétrés de ces deux questions et plus nous serons forts vis-à-vis de notre adversaire. Plus nous serons nous-mêmes. A bien considérer les choses, la lutte entre l'Occident et nous n'a jamais cessé. Il n'y a pas même eu de trêve, il n'y a eu que des intermittences de combat. Maintenant à quoi bon se le dissimules? cette lutte est sur le point de se rallumer plus ardente que jamais et cette fois encore comme autrefois, comme toujours c'est l'église de Rome, l'église latine qui est à l'avante-garde de l'ennemi.

Eh bien, acceptons le combat, franchement, résolument. Qu'en face de Rome, l'église d'Orient n'oublie pas un seul instant qu'elle est l'héritière légitime de l'église universelle.

... Le catholicisme a de tout temps fait toute la force du Papisme, comme le Papisme fait toute la faiblesse du catholicisme.

La force sans faiblesse n'est que dans l'église universelle. Qu'elle se montre, qu'elle intervienne dans le débat et l'on verra de nos jours ce qu'on a déjà vu dans les trois premiers jours de la réformation, alors que les chefs de ce mouvement religieux qui avaient déjà rompu avec le siège de Rome, mais qui hésitaient encore à rompre avec les traditions de l'Église Catholique, en appelaient unanimement à l'Église d'Orient. Maintenant comme alors la réconciliation religieuse ne peut venir que d'elle; elle porte dans son sein l'avenit chrétien.

Telle est la première, la plus haute question que nous ayons à debattre avec l'Europe occidentale, c'est la question vitale par excellence.

Il y en a une autre bien grave aussi; c'est celle que l'on appelle communément la question d'Orient; c'est la question de l'Empire. Que l'histoire nous dit?

... Elle nous dit que l'Orient orthodoxe, tout ce monde immense qui relève de la croix grecque, est un dans son principe, étroitement solidaire dans toutes ses parties, vivant de sa vie propre, originale, indestructible. Il peut être matériellement fractionné, moralement il sera toujours un et indivisible. Il a subi momentanément la domination latine, il a subi pendant de siécles l'invasion des races asiatiques, il n'a jamais accepté ni l'une ni l'autre.

... Qui ne voit que l'Occident avec toute sa philanthropie, avec son prétendu respect pour le droit des nationalités et tout en se déchaînant contre l'ambition insatiable de la Russie, ne voit dans les populations qui habitent la Turquie qu'une seule chose: un proie à dépécer.

... Mais cette Providence historique qui est au fond des choses humaines y a heureusement pourvu. Déjà au treiziéme siècle l'Empire d'Orient tout mutilé, tout enervé qu'il etait a trouvé en lui-même, assez de vie pour rejeter de son sein la domination latine après soixante et quelques années d'une éxistence contestée; et certes il faut convenir que depuis fors, le véritable Empire d'Orient, l'Empire orthodoxe, s'est grandement relevé de sa déchéance.

... L'immense question sera décidée ... ici, parmi nous, au centre, au coeur même de ce monde de l'Orient Chrétien, de l'Orient Européen que nous réprésentons, de ce monde qui est nous-même. Les destinées définitives qui sont aussi les nôtres, ne dépendent que de nous; elles dépendent avant tout du sentiment plus ou moins énergique qui nous lie qui nous identifie l'un à l'autre.

... Encore une fois ce qui fait le fond de cette hostilité, ce qui vient en aide à la malveillance qu'ils exploitent contre nous, c'est cette opinion absurde et pourtant si générale que tout en reconnaissant, en s'éxagérant peut-être nos forces matérielles, on en est encore à se demander si toute cette puissance est animée d'une vie morale, d'une vie historique qui lui soit propre. On, l'homme est ainsi fait, surtout l'homme de notre époque, qu'il ne se résigne à la puissance physique qu'en raison de la grandeur morale qu'il y voit attaché.

... Voilà un Empire qui par une recontre sans exemple peut-être dans l'histoire du monde, se trouve á lui seul représenter deux choses immenses: les destinées d'une race tout entière et la meilleure, la plus sainte moitié de l'Église Chrétienne.

Et il y a encore des gens qui se demandent sérieusement quels sont les titres de cet Empire, quelle est sa place légitime dans le monde!.. Serait-ce que la génération contemporaine est encore tellement perdue dans l'ombre de la montagne qu'elle a de la peine à en apercevoir le sommet?....

... Est-il étonnant que de vieilles convictions luttent de tout leur pouvoir contre une évidence qui les ébranle, qui les suprime? et ne serait-ce pas à nous à venir en aide à cette évidence, à la rendre invincible, inèvitable?"

6. In the original:

"...homme intelligent, doué d'energiques sentiments de nationalité, profondément dévoué au service de l'Empereur et qui par une longue expérience de la presse aurait acquis une connaissance suffisante du terrain sur lequel il serait appelé à agir.

...Si cette idée était agrée, je m'estimerais trop hereux de mettre aux pieds de l'Empereur tout ce qu'un homme peut offrir et promettre: la propreté de l'intention et le zèle du dévouement le plus absolu."

7. For more details, see Ospovat 1992.

Works Cited

Aksakov, I. S. 1886. *Biografiia Fedora Tiutcheva*. Moscow.

Azadovskii, K. M., and A. L. Ospovat 1989. "Tiutchev v dnevnike A. I. Turgeneva." *Literaturnoe nasledstvo*, 97, #2. 63–98.

Diary of George Mifflin Dallas, United States Minister to Russia; 1837–1839. 1970. New York.

Kazanovich, E. P. 1928. "Iz miunkhenskikh vstrech F. I. Tiutcheva." *Uraniia; Tiutchevskii al'manakh 1839–1839*. Leningrad. 125–171.

Kuzina, L. N. 1989. "Tiutchev v dnevnike i vospominaniiakh M. P. Pogodina." *Literaturnoe nasledstvo*, 97, #2. 7–27.

Lane, R. 1987. "F. I. Tyutchev's Service Absenteeism and Second Marriage in the Light of Unpublished Documents; (1839)." *Irish Slavonic Studies* 8: 6–13.

Ospovat, A. L. 1992. "Tiutchev i zagranichnaia sluzhba III Otdeleniia (materialy k teme)." *Tynianovskii sbornik: piatye Tynianovskie chteniia*. Riga. [forthcoming]

___. 1985. "Tiutchev letom 1837 goda." *Literaturnyi protsess i razvitie russkoi kul'tury XVIII-XIX vv*. Tallinn.

Pigarev, K. 1962. *Zhizn' i tvorchestvo Tiutcheva*. Moscow.

"Pis'ma F. I. Tiutcheva k ego vtoroi zhene." *Starina i novizna*. 1914. vol. 18.

Tiutchev, F. I. 1988. "Nezavershennyi traktat 'Rossiia i Zapad.'" *Literaturnoe nasledstvo*. 97, #1. 183–300.

___. 1913. *Polnoe sobranie sochinenii*. St. Petersburg.

___. 1984. *Sochineniia v dvukh tomakh*. Moscow.

HISTORIA SUB SPECIE SEMIOTICAE

The Rhetoric of Representation of Political Leaders in Soviet Culture

MIKHAIL YAMPOLSKY[1]

New York University

THIS ARTICLE is devoted to the rhetoric of political representation in Soviet culture. Or, more precisely, the rhetoric of representation of political leaders, particularly Lenin and Stalin. I believe that analysis of images representing authority can provide valuable insights into the nature of that authority, in its 'unconscious' self-determination.

Political representation is in fact doubly representational. Any representation consists in a substitution, doubling, signification of one thing through another. But in the case of political authority representation is placed at the very heart of its social functioning. Authority legitimizes itself by **representing** something, such as the people, a social class, a political party, the voters, and so on. Consequently, on the one hand, the image of a leader is a representation of his physical persona; and on the other hand, it is a representation of a leader as a representative, as a body already imbued with signification and symbolism. This ambivalence is specific to political representation—a quality that organizes and predetermines its semiotic structure. An official portrait is not "simply" a portrait of an individual, it is a portrait of a portrait. It eliminates all forms of movement, as something destructive for a tautological representation. Movement is action, a transition from one state into another; consequently, it interferes with the "pure" representationality of a portrait. A monarch in a portrait cannot move. The increase of his 'statuesqueness' is proportionate to the increase of his/her authority. Thus Napoleon, for instance, as a revolutionary general and then as a consul can be depicted in a dynamic form. Napoleon the emperor, however, is always motionless. Something similar to this can be detected in representations of Stalin, whose motionlessness increases with the years and is

Elementa, 1993
Vol. 1, pp. 101–113
Reprints available directly from the publisher
Photocopying permitted by licence only

symbolized by his overcoat, which gradually extends lower, covering his lower legs and finally transforming the leader into a monumental pole.

Tautology, self-referentiality, assertion of representation as such, as the main essence of representation, makes the rhetoric of political imagery extremely conservative. In many cases it can be traced to comparable Indo-European archetypes.

The conservatism of political representation often contradicts the ideology and the politics of the represented authority. Particularly strange phenomena can be observed when political representation develops under circumstances which are exterior to the national traditions. In discussing Russia we need to take into consideration the fact that Russian tradition provides no *topoi* of representation for revolutionary and democratic leaders. Over the ages it merely reproduced rhetorical cliches of religious and monarchist political representation. Rudimentary elements of this tradition could not help but enter into the arsenal of post-revolutionary rhetoric. To understand the paradoxes of this rhetoric, we need to turn to that tradition in which it is grounded, and which it also polemicized against.

Ernst Robert Curtius has revealed the existence of a classical dichotomy, possessing a universal character, which is significant for the depiction of the ruler. The dichotomy is found in the opposition of qualities usually assigned to the ruler: the rage of the warrior against the wisdom of the statesman.[2] The Indo-European roots of this opposition were revealed before Curtius by Georges Dumézil, who noted that in the Indo-European pantheon of gods there is an inseparable dyad, *Mitra* and *Varuna*, who have opposite but complementary qualities: Mitra is benevolent, kind, and sympathetic, while at the same time Varuna is vindictive, fearsome, and menacing.[3] Dumézil has shown that these polar but paired qualities in time transferred to the archetypal dyad of rulers who succeed Mitra and Varuna, the Roman rulers *Romulus* and *Numa*. The menacing conqueror Romulus turns out to be a politicized version of Varuna and the philosopher Numa is nothing more than Mitra. The development of monarchial political representation eventually leads to the unification, in the solitary figure of the monarch, of both sides of the opposition. The representation of the political rhetoric of the monarchy embodies characteristics of both Mitra and Varuna.

This equilibrium of characteristics of Mitra and Varuna was upset at the moment of formation of the Western European democracies. In this new system of political representation, the point was reached where the solitary state of wiseman and warrior was split in two and transferred to different figures, the military commander and the legislator.

As Henri Focillon notes, French kings in the period of Absolutism still preserved, in their everyday life, features or traits of their ancient ancestors, the Gothic rulers. Like the Goths, they spent their time making war and hunting.[4] The passion of European monarchs for hunting, and the countless hunting mansions of kings and dukes throughout all of Europe, are not simply the memory of a Gothic past. In the system of the representation of the sovereign, this is the sign of Varuna, a warrior and hunter. Not accidentally, portrayal of the king hunting remained the permanent motif of official painting. With the appearance of the bourgeois democracies, hunting continued to be an important feature of the feudal aristocracy, but it completely disappeared from everyday life of the new type of politician, who appeared first of all as the legislator and the representative of the people, taking on characteristics of Mitra—Numa, i.e., the wiseman.

How is the legislator to be depicted in the new system of representation which is forming at the end of the eighteenth century? Rousseau describes the functioning of democratic authority as follows: "The ruler who has no other power but a legislative one functions exclusively with the help of laws; and because laws are truly the acts of a common will, so the ruler can function only in the presence of an assembled people."[5] Hence, the leader functions only in the presence of all people, or the representatives of the people in the assembly, and the meaning of his function becomes the *word* in which the *law* is formulated. From this follows two fundamental qualities of a democratic ruler—oratorical skill (the ability to use *words*) and absolute objectivity (allegiance to the *law*).

In Jean-Louis Laneuville's typical portrait of the republican legislator Barer, entitled *Barer, demanding the execution of Louis XVI at the Convention* (*Barer demandant à la Convention la condamnation à mort de Louis XVI*), these fundamental elements of new political representation are expressed with maximum completeness. We see a middle-aged man. He leans on the podium with his right hand. On the edge of the podium there are papers with a written speech. His gaze is fixed on the audience. There are only two attributes of power near Barer—the podium, site of the materialization of the word, and the papers, a sign of the law. J. Davallon, who has analyzed this portrait, has shown how this combination of podium and paper led to a fulfillment of the symbolic joining of word and law. He notes: "the word substantiates the writing."[6]

The podium becomes the sign of the democratic leader, and beginning with the French Revolution it is ceaselessly reproduced. When the sculptor L. Kerbel created the statue of Marx in Moscow he placed the founder of scientific communism at the podium. In this case, we can talk of a pure *topos*, one that does not reflect any political or biographical realities.

Lenin[7] is constantly depicted behind the podium. In such an obviously rhetorical text as Dziga Vertov's film *Three Songs of Lenin*, a political portrait of Lenin is constructed as a series of images of him at the podium, interrupted by intertitles such as "Here he is speaking passionately, inspiring the masses..."; "The founder of the Communist International..."; "The leader of world communism..."—and so on. However, early on Lenin's podium takes on whimsical new forms. One of the first symbolic variants is created by El Lissitzky in 1924. He designs a podium as a gigantic construction upon which Lenin's figure is situated so high that the podium loses all functional meaning. It is impossible to address the people from such a distance. It is interesting to note that one of the issues of *Soyuzkinozhurnal* (a serial newsreel) in that same year 1924 included a section entitled *The Parade of the Balloons* in which the "podium in the sky" was represented. This sky podium—the basket of a balloon where the speaker was placed (with the papers of his prepared speech), ascended to such a high position that he could not be heard.

The final transformation of this topic is the image of Lenin on top of a mountain (engraving by P. Staronosov) in the midst of the natural Sublime, where the audience disappears, the speech disappears and what remains is only the romantic reign over the limitless vastness. In part, this aloofness is connected to the symbolism of the orator's voice. The meaning of the orator's voice in the eighteenth century (specifically, the voice of William Pitt) was formulated by Peter de Bolla as follows: "A voice sounds out across the nation, a voice of the nation sounds out across the waves, a voice so sublime, so powerful it causes distant thrones to shudder."[8] This mystical voice of freedom is hovering so high that it is transformed either into an inarticulate thunder or into sublime silence.

Here we have a *topos* of political representation which has undergone amazing transformations. Barer in Laneuville's painting was situated at eye-level, and appeared as a symbolic citizen equal to other citizens. His power was concentrated in the audible word. The aerial podiums of El Lissitzky and the film chronicle of 1924 radically alter the meaning of the situation. Here the leader is no longer the representative of the people, but an icon which mystically ascends into the sky. His word loses its connection with law and assembly and turns into the divine word, not intended for the subjects. Formally, we still have the sign of a democratic leader, but he legitimates his power not through word and law, but through his ascended position above the crowd.

Such a transformation of *topoi* or rather their strange hybridization is typical for the post-October period. Thus, for example, features of a feudal noble are blended with the image of a democratic leader. Thus occurs a peculiar renascence of the Varuna state. In particular, this happens by means of the *topos* of hunting, mentioned above. Right after Lenin's death, in the mid-twenties, a remarkable theme begins to proliferate: "Lenin Hunting." The new theme is reflected in the prose of N. Krylenko, A. Polikashin, M. Prishvin, and others. Basically, we have here, projected upon Lenin, traits of the system of the representation of power which so recently dominated Russia—the autocracy.

In Prishvin's collection of stories, *Lenin Hunting* (1926), the leader behaves strangely, failing to shoot at the bird. The writer clearly feels this "strangeness" in the situation he reveals to the reader. He finds it is necessary to include in the story a political motivation for Lenin's love of hunting:

> Every hunter is a great individualist because each wants to train his dog better and shoot better than his fellow hunter. But deep in the soul of every real hunter who learned this passionate habit in the days of his childhood, some spontaneous communism lives on. This is the only reason why, in hunting, people from very different social stratas can greet one another as friends.[9]

A predilection for hunting is part of the lifestyle of the subsequent Soviet leaders from Brezhnev to Yeltsin. The only exception to this rule was Gorbachev—the only Soviet leader who was oriented in his representations towards Western stereotypes, which, in part, is the reason for his popularity abroad and for the lack of his popularity in the U.S.S.R.

Rhetorical oxymorons arise in the post-October period as a reflection of the contradictory status of the new state—on the one hand it is a dictatorship, on the other, a democracy. This combination of the irreconcilable begets a certain whimsicalness in the new emblematics.

One example of this unrepictibility in symbolism is another metamorphosis of the podium. One of the most bizarre transformations of this *topos* is the image of Lenin standing on the armored car, known to us through S. Eisenstein's film *October* and the statue in Leningrad. Turning the podium into an armored car demonstrates how the place of the realization of the word (i.e., of democracy) turns into the instrument of dictatorship. The orator on the armored car is a combination of Varuna and Mitra which is much more bizarre than any fantasy of the emblematic system of the monarchy.

It is revealing that in another political system which was also oriented on the "combination" of dictatorship and democracy, that of fascist Italy, we find exact replicas of the same rhetoric. Mussolini gave one of his speeches astride a tank. Roberto Rossellini recalls that in Ladispoli, where he lived, one fascist functionary, imitating Mussolini, gave a speech astride a tractor.[10] This substitution of tractor for tank testifies to the fact that here we have a powerful mechanism symbolically transferring part of its strength to the orator. Yuri Annenkov recalled that Trotsky, for his parade portrait, chose the leather suit of the chauffeur (the prototype for the favored leather jackets of the revolutionaries). Setting himself in his portrait dressed in leather, Trotsky gave, in the subtext of his depiction, the car as a kind of invisible yet implicit podium. It is worthwhile to remember Yeltsin's speech delivered while standing on a tank during the August 1991 coup attempt. Doubtlessly, Yeltsin extensively uses the rhetorical repertoire of Gorbachev's predecessors.

Stalin was never an orator, although he is often depicted standing on podiums. A steel podium is not as much part of his image as 'steel' became a conscious root of his pseudonym. Yet, for instance in M. Avilov's painting from 1933, *Stalin's visit to the First Cavalry Army* Stalin is represented on a sled-podium, an essential parody of an armored vehicle.

The relationship in Soviet political rhetoric between a podium and the democratic tradition gradually dwindles. Instead the podium becomes connected to a pulpit of a church preacher. As Michael Fried has noted, it is around the performance of a preacher (St. Augustine) that a new poetics is formulated in the paintings of Carl van Loo in 1750s. This poetics Fried calls "absorption."[11] In that case the orator was the center of hypnotic fascination of the audience. The auditorium in the painting becomes a kind of a mirror, in which the unique, superhuman status of the orator is reflected. For Van Loo it is a saint. The speech becomes a kind of miracle, a revelation. By the same token, the transfer of attention to the response of the audience transforms the orator into a personified void, and confirms him as a pure representation of a rhetorical effect aimed at the audience. This is the type of depiction of an orator that dominates Soviet representation. Both Lenin and Stalin are surrounded by the attentive public. In this context, the orator becomes a prophet.

This quasi-religious meaning of a podium explains its own metamorphoses. Its transformation into a moving mechanism in reality reminds of the archaic types of the emperor's throne, of the divine throne, the so-called *chariot throne*, which was used by rulers or gods for ascending to heaven and assuming the 'rank' of cosmocrats—of the rulers of the universe.

As H. P. L'Orange has shown, gesture, particularly characteristic of depictions of Soviet leaders, belongs to the cosmocratic repertoire. This is the raising of the right (in certain occurrences of the left) hand--a gesture reminiscent of Eastern solar divinities, and one that becomes a symbol of the emperor's authority with Constantine. In Christian iconography this gesture belongs to Christ-Pantocrator. At the same time it is a symbol of the omnipotence of Jahve which was often described in the Bible. The psalmist sings: "Thy arm is with might. Let thy hand be strengthened, and thy right hand exalted" (Psalm 88: 14, *Douay Rheims Version*). Omnipotence itself is manifested in Jahve's outstreched right hand: "I made the earth, and the men, and the beasts that are upon the face of the earth—says Jahve—by my great power, and by my streched out arm" (Jer. 27: 5; 32: 17).[12]

Of course, this is not to suggest that the iconography of leaders in the U.S.S.R. consciously imitated the gestures of God, emperor, or Christ. But there is some internal consistency in the fact that among a variety of gestures in post-revolutionary Russia the leaders are depicted with the one that reminds of the familiar representations on icons and imperial effigies.

This is instructive because Stalin, in his iconographic development, follows a trajectory similar to that of Christ-Pantocrator. The gesture of blessing, the so-called *Benedictio Latina* , was prior to this signification a gesture of speech which was widely popular in the ancient East. This gesture—gestus oratorius—accompanied Christ in the depictions of divine disputes—divina disputatio, sacra conversazione, and only later was separated from the word and became a gesture of blessing and authority.[13]

Something similar can be detected in Stalin, who gradually deviates from the post-revolutionary model of the representation of the leader and develops his own specific system. For the most part, Stalin rarely uses the *topoi* of democratic power. The podium is not a part of his arsenal; he speaks infrequently and reluctantly. However, there is an oxymoron which is very characteristic for him—that is, his depiction in military uniform behind a writing desk. Stalin prefers to portray himself as a militarized wise man. But this kind of rhetoric sends us back to the archaic feudal emblematics as well. In the message of the Gothic king Vitigis, written by Cassiodorus in the sixth century, we could read: "And not in the cramped chamber but in the wide-open battlefield I was chosen, and you must realize this."[14] Here we have the idea of *choosing* the monarch projected onto his martial deeds. The monarch does not take his power by force of arms, but rather demonstrates his strength to God, who then chooses him on the battlefield. The military uniform of Stalin appears to be a sign of such designation on the battlefield of the past (to take part in revolutionary battles was a form of legitimization for many Soviet leaders of the "heroic" period).

On the more visible level Stalin imitates the imperial attitude of the 18th to 19th centuries, when the custom of European monarchs was to appear almost always in some regimental uniform, or in that of an army general. The military uniform makes an appearance in Europe in the late 17th century. According to Ernst Kantorowicz, this moment corresponds to the time "when the continental nations asserted themselves as military monarchies, with a prevalence of the military at large."[15] The adoption of a uniform by Stalin probably reflects a similar shift in the status of the Soviet state. This fact also reflects the loss of "heroic" orientation among the Soviet ruling class, and a new preference for a soldier.

Kantorowicz also recalls that in late antiquity gods are represented in a uniform; he provides some examples of Christ depicted in a military attire (the so-called *Christus miles*), sometimes in an exact replica of a uniform of a Roman general. Curiously these images represent the *imitatio imperatorum* on the part of the gods, a tendency opposite to that which existed before—an *imitatio deorum* on the part of the emperors.[16] Stalin's gesture of adoption of the uniform contains some strange remnants of this late Roman phenomenon.

But the military uniform is not enough for legitimization, and Stalin also includes in his representation his Kremlin office—on the one hand connected as a *topos* close to the image of Lenin, on the other hand as the rhetorical place of meditation and wisdom. In Stalin's emblematic, thinking replaces oratory.

André Bazin, analyzing Stalinist war films, notes that during the time of the bloodiest battles, the leader remains in the quietude of his office, where there was a "submersion into a thoughtful and semi-hermitical atmosphere," which provides an "urge toward scientific studies." Bazin says: "Stalin in loneliness, pondering the map, and after a long but intensive meditation, and a few puffs of the pipe, alone decides the fate of the military operation."[17]

Bazin rightfully noted the significance of a geographical map in the new system of representation. In 1928 Walter Benjamin during his stay in Moscow noticed the great profusion of these maps. In its own way a map symbolizes a victory of space over time, the final end of history and its replacement by geography. History is now made by the leader, who shapes the Earth. There is a poster where Lenin and Stalin are drafting maps (V. Govorkov: *In the Name of Communism*, 1951). Lenin here is armed only with one map, whereas Stalin has three. In this poster Stalin erases the word 'desert' from the map. This of course is not a gesture of a democratic leader but of a divine demiurge. In V. Oreshnikov's painting *Lenin and Stalin in the Headquarters of Petrograd's Defence* (1949) the distribution of roles is very characteristic. Stalin is standing behind Lenin by the map. It is towards him that everyone's gaze is directed. The one who places his hand on the map, that is, on the "space", is the one who possesses the ultimate authority.

The map is a particular variant of the 'writing of power.' Gradually writing becomes more and more significant and takes up the place of the word. The writing leader becomes more important than the speaking one. He no longer directs his word to the people, but rather, in solitude and silence, he writes the new law. A paper scroll becomes a major attribute of the leader—not a notebook, not a sheet of paper, but a scroll which provides the image with archaic symbolism. Lenin is represented with a scroll in his hand in many statues: by A. Kibalnikov (1950), P. Yatsenko (1950), N. Tomsky (1974) and so on. Stalin's effigies, obviously imitating the portraiture of the father-founder and made by the same artists (Kibalnikov 1950; Tomsky 1951) at about the same time repeat this gesture. In the iconography of Christ-Pantocrator, which belongs to the so-called *traditio legis*, Christ is often represented with a scroll containing the law standing on the top of the Earth with his right hand raised. Also the scroll of the law as a symbol of authority can be found in many official portraits, for instance in Velazquez.

The increased emphasis on writing as opposed to the spoken word is indicative of a new representational situation which characterizes Stalin. Silence, quietude, and loneliness indicate the peculiarity of Stalin's attitude towards the democratic political regime. Stalin receives power not from the people's representatives, but from his predecessor—the sacred leader. The symbolic act of receiving power was fixed in Stalin's case by taking an oath before the coffin of Lenin. The system of political representation provides this act with a special meaning. In Mikhail Chiaureli's film *Oath* (1946), Stalin offers his oath of faithfulness to Lenin before the grave of the leader, surrounded by masses of people. However, he addresses his words not to the people but to the deceased.

The oath is an extremely archaic act of legitimization for the sovereign. It takes on special meaning in revolutionary periods and, as Jean Starobinski notes, it realizes itself in symbolic opposition towards the ceremony of the monarch's coronation: "Revolutionary oath *creates* sovereignty, while the monarch *receives* it from heaven."[18] The oath, it should be noted, is the most theatricalized form of contract. The classical oath of revolution is the ritual to be realized before the face of Law. In 1789 George Washington gave an oath to the American Constitution and in the same year representatives of the third estate in Paris, on the tennis court, gave the famous oath which signified the creation of the National Assembly.

The oath of Stalin seemingly reproduces revolutionary patterns, transforming them again. The new leader gives an oath not before an assembly of citizens, nor before the face of Law, but before the face of the deceased leader, from whom he inherits his power. In this act of legitimization the monarchic *topoi* blend with the rhetoric of democracy. At the same time, Stalin's oath constitutes the cult of Lenin's remains, which provides Stalin with the full weight of his power and legitimacy. This is why the mausoleum appeared as a sacred place for Stalin and his successors.

There are three versions of this act which legitimize Stalin's authority. First there is a drawing by P. Vasiliev from 1937. Here Stalin is represented on a podium of the Second Congress of the Soviets of the U.S.S.R. on January 26, 1924. The oath is given here in front of the People's legislating Assembly. However, the meaning of the event is conveyed due to the explicit duplication of Stalin's figure by Lenin's portrait in the back. The importance of the direct replication of the angle in which Stalin's head is turned by Lenin's portrait is that it symbolically transforms Stalin into the living embodiment of the deceased Lenin, and perhaps more importantly, into a repetition of a representation.

The second scene of the oath was produced in 1950–51 by the painter M. Antonian. Here Stalin swears next to Lenin's coffin. The Assembly, the convention are deleted, for they are no longer needed. But perhaps the most canonical model can be found in the relief *Oath* made in 1949 by E. Vutechich and a group of sculptors. This relief was obviously made on the model used in Chiaureli's film. Here Stalin swears from a podium which is located at the same spot on which Lenin's mausoleum will be erected. The sacred place of the legitimation of Stalin's authority also constitutes the enunciation of the eternal tomb of the leader. The mausoleum becomes the main temple of Stalin's authority. In all three of the images Stalin places his right hand on his chest. On the one hand, it is a repetition of Lenin's famous gesture, and on the other, it is a symbol of accepting an oath which will be subsequently repeated in thousands of Stalin's effigies.

From this it follows—one more transformation of the *topos* of the podium—the construction of the podium on Lenin's mausoleum. If Lenin receives his power on the podium-armored car, so Stalin receives his on the podium-tomb of Lenin. Basically, this primary symbolic podium of the country is completely divorced from the word. This is the podium of silence, the podium of the silent communion of the country's leaders with the remains of the founding father. This podium is not intended for words or written texts. The text realizes itself before the podium in the form of writings and images passing before it, carried by the masses. Basically, the very masses themselves turn into texts which are read from above by a podium who remains in silence.

In this context one may recall a painting by I. Davidovich and E. Tikhanovich *Glory to the Great Stalin!* (1950), where the solemn and speechless masses are simply exposed to the gaze of the leader on the top of the mausoleum. The painting is structured in such a way that the figure of Stalin appears precisely above the inscription "Lenin." Stalin literally becomes the reincarnation of Lenin.

The doubling of the two figures of authority turns into a categorical canon. Starting with the 1930s the figures of Lenin and Stalin are depicted simultaneously in most paintings. In those paintings which represent Lenin alive as a rule Stalin stands behind him, duplicating by his profile the profile of the founding father. After Lenin's death the situation is inverted: now behind Stalin's back appears the sculptural figure of the founding father. For example, in a poster by Iraklii Toidze, *With the Banner of Lenin, Under the Leadership of Stalin, straight to the Victory of Communism!*, Stalin stands on a podium raising his right hand. Behind him appears a statue of Lenin making an exactly similar gesture. These depictions are numerous. Sometimes the situation becomes grotesque. On the medallion of M. Manizer the two leaders are "glued" together and turned into a unified body. The two bodies have only two hands which reproduce one symbolic gesture. Stalin holds the scroll of law with which we are already familiar; Lenin keeps his right hand on his chest.

This two-headed and two-handed monster exemplifies the situation of representational doubling, which, as was noted above, lies at the basis of any political representation. The living is duplicated by a representation of the dead—a portrait, or, more often, a statue. This doubling blurs the boundaries between the figures, between the living body and its representation. The reigning leader achieves his final metamorphosis into an icon, a tautological symbolic doubling of the self. Here representation is already completely severed from its democratic models. The assembly of the people can no longer 'fit' into the narrow space circumscribed in its entirety by the two figures. Representation, fully subordinated to doubling, becomes representation of absolute authority par excellence, where there is no space that separates the symbolic body of the monarch from God.

The paradoxical contradictions of Soviet political *topoi*, connect with rudimentary elements of the tradition of Russian monarchic representation, as well as with the extreme contradictions inside the U.S.S.R.'s social system, which combines democratic rhetoric with totalitarian practice. This is why *topoi* of political representation in the U.S.S.R. could be included in the sphere of some kind of rhetorical *teratology*. However, this teratologicalness, this oxymoroness, this combination of the irreconcilable has fundamental ideological functions.

For Soviet ideology of the totalitarian period, the urge to fulfill what cannot be fulfilled, or the urge towards superhuman goals, which lie beyond the limitations of human possibilities, is extremely characteristic. As justly noted by Hannah Arendt, a totalitarian movement, for the sake of its own permanent existence as *a movement* must constantly violate social stability.[19] This is why it constantly brings to the ideological picture of the world the element of contradiction, the struggle of opposites. The absurd is introduced into ideological representation as a form of negation of reality, and as evidence of the unlimited power of totalitarian society, of totalitarian consciousness. Teratological rhetoric of political representation merely reflects this urge of ideology towards the assimilation of the contradictions and absurdities which are transformed by this ideology into *normal*, conventional. This is why the rhetorical oxymoron is, relatively speaking, painlessly transformed into an icon.

Notes

1. The first version of this text was presented as a paper at the conference *Representation of State: State of Representation* at New York University (1990). A. Maksimov, D. Stirk and A. Waintrub contributed to its translation into English.
2. E. R. Curtius, 1953. *European Literature and the Latin Middle Ages*. London, 1953, pp.167–182.
3. G. Dumézil, *Mitra-Varuna. Essai sur deux représentations indo-européennes de la souveraineté*. Paris, 1948. p. 62. ["...là tumulte, passion, impérialisme d'un *iunior* déchaîné; ici, sérénité, exactitude, modération d'un *senior* sacerdotal."]
4. H. Focillon, *L'an mil*. Paris, 1984, p.19.
5. J.-J. Rousseau, *Du contrat social*, Paris, s.d., p.98.
6. J. Davallon, *Représenter le législateur: portrait du citoyen ou effigie du héros*, Procès, 1983, n 11–12, p.126.
7. For a general study of Lenin's representation in Soviet political culture see: N. Tumarkin, *Lenin lives!*. Cambridge, Mass., 1983.
8. P. de Bolla, *The Discourse of the Sublime: Readings in History, Aesthetics and the Subject*, New York, 1989, p.143.
9. M. Prishvin, *Izbrannoe*, Moscow, 1971, p.155.
10. R. Rossellini, *Fragments d'une autobiographie*. Paris, 1987, p.69.
11. M. Fried, *Absorption and Theatricality: Painting and Beholder in the Age of Diderot*, Berkeley–Los Angeles–London, 1980, pp.17–27.
12. H. P. L'Orange, *Studies on the Iconography of Cosmic Kingship in the Ancient World*, New Rochelle–New York, 1982, pp. 139–170.
13. H. P. L'Orange, pp. 171–197.

14. Cited in: V. Ukolova, *Antichnoe nasledie i kul'tura rannego srednevekov'ia*, Moscow, 1989, p. 106.
15. E. H. Kantorowicz, *Selected Studies*, New York, 1965, p.14.
16. E. H. Kantorowicz, p. 19–21.
17. A. Bazin, "Le mythe de Staline dans le cinéma soviétique"—In: *Qu'est-ce que le cinéma?* t. 1. Paris, 1958, p. 80.
18. J. Starobinski, *1789. Les Emblèmes de la raison*, Paris, 1979, p.66.
19. H. Arendt, *The Origins of Totalitarianism*, San Diego–New York-London, 1979, pp. 391–392.

NOTES FOR CONTRIBUTORS

Submission of Articles

Articles may be submitted to either Vyacheslav Ivanov, Editor, or Alexander Ospovat, Associate Editor, at the journal's office: Department of Slavic Languages and Literatures, U.C.L.A., 115 Kinsey Hall, 405 Hilgard Avenue, Los Angeles, California, 90024, USA. Authors are asked to submit *three copies* of their articles to facilitate and speed the review process; they will be informed of the decision at the earliest possible date. Submissions to the journal will be evaluated by the editors and members of the editorial advisory board; outside specialists may be consulted, as needed.

Submission of an article will be taken to imply that it has never been published previously in any form and has not simultaneously been offered to any other publication. It is a condition of acceptance of an article that the publisher acquires the copyright of the typescript. Therefore the article cannot be published elsewhere in the same form, in any language, without the publisher's consent.

Form and Length

All articles should be typed on one side of the paper only, and double spaced with a wide (3 cm) margin on each side of the text. Pages of each copy must be sequentially numbered. Section headings should be entirely in capitals, on a separate sheet of paper; titles should be in small letters, with main words capitalized. In either case, the text should start on the next line. The maximum preferred length is twenty double-spaced pages; there is no minimum length. Exceptionally long articles will be considered when of particular merit, and short notices are welcome.

Names, affiliations and complete mailing addresses for authors should appear on a separate title page. If affilitations for authors differ, list addresses separately beneath each author's name. Indicate for the typesetter which author will check proofs. Authors should also provide an abbreviation of the paper's title (no more than 35 characters) for use as a running head. Include acknowledgements under a separate head at the end of the paper but before the Works Cited list.

The publisher encourages authors to submit accepted manuscripts on computer disks. Word Perfect 5.1 is the preferred software, but other software and formats are acceptable, as are all sized disks. Authors must enclose a printed copy of the manuscript (in triplicate) along with the disk. All disks should be marked with the name of the software package that was used and the file name. Disks will be returned to the author with page proofs.

Illustrations

Illustrations should be presented as "camera-ready copy," numbered with consecutive arabic numbers (Figure 1, Figure 2, etc.), have descriptive captions, and be mentioned in sequential order in the text. Keep illustrations separate from the text, but indicate clearly their approximate position within the text. Permission to reproduce illustrations and any other material which has been previously published must be obtained by the author.

Preparation: Line drawings should be prepared in black (India) ink on white paper or tracing cloth, with all necessary lettering completed. Photographs must be good original prints of maximum contrast. Color prints or transparencies should be submitted only if the color is necessary to the understanding of the discussion.

Captions: A list of captions relating to the illustrations, with relevant numbers (Figure 1, Figure 2, etc.), should be set out on a separate sheet of paper attached to the typescript. The permission statement provided by the copyright holder should be included at the end of each caption.

References

We encourage the use of the author-date system of documentation. Within the text, references are indicated either as: "Recent work (Konrad 1972: 125–127)" or "Recently Lotman (1983) has shown. . . ". The full list of references, under the heading "Works Cited," should be collected at the end of the paper in alphabetical order, and set out in the manner described and illustrated below. Note that the initials of the first author (only) are placed after the name.

Books

Author's name. Year. Book title (underlined). Location: Publisher. Chapter or page numbers.

Journal Articles

Author's name. Year. Article title. Name of journal (underlined), volume number: inclusive page numbers.

Edited Collections

Author's name. Year. Article or chapter title. 'In': followed by book title (underlined). (Ed.) Editor's name, inclusive page numbers. Volume number. Location: Publisher.

Examples:

Lerdahl, F. and R. Jackendorff. 1983. A Generative Theory of Tonal Music. Cambridge, MA and London: MIT Press.

Cassedy, S. 1991. "Pavel Florensky's philosophy of language: its contextuality and its context." Slavic and East European journal 35: 537–552.

Andersen, H. 1991. "On the projection of equivalence relations into syntagms." In: New Vistas in Grammar: Invariance and Variation. Eds. L. R. Waugh and S. Rudy, 287–311. Current Issues in Linguistic Theory 49. Amsterdam and Philadelphia: John Benjamin.

Notes

Notes should be substantive in nature; they are to be gathered together and placed following the body of the article.

Proofs

Authors will receive page proofs (including illustrations) by airmail for correction, which must be returned to the printer within 48 hours of receipt. Please ensure that a full postal address is given on the first page of the manuscript so that proofs will arrive without delay. Authors' alterations over 10% of the original composition cost will be charged to authors.

Reprints

The publisher will supply 25 free reprints of each paper to the first named author. Further reprints may be ordered by completing the appropriate form sent with proofs. Free reprints are sent by surface mail, but ordered reprints by airmail. The two sets will not therefore normally arrive together.

Forthcoming Articles . . .

ELEMENTA

History, Meaning and Definition of the Term Semiotics
VYACHESLAV IVANOV

The Girl on the Hill: Parallel Structures in *Pride and Prejudice* and *Eugene Onegin*
RICHARD TEMPEST

Sound as Vision in Lermontov's "Vyxožu odin ja na dorogu"
NANCY POLLAK

Text and Subtext of the Hungarian Renaissance: Matthias Corvinus and the Lesser Nobility
MARIANNA BIRNBAUM

The Double-Faith in "The Song of Igor's Campaign" and the Theme of the City
TATIANA NIKOLAYEVA

Pasternak, Zamiatin and Bradshaw
OMRY RONEN

Nabokov and the Third-Rate Writers (On the Sources of *Lolita*)
ALEXANDER DOLININ